RADON
THE INVISIBLE THREAT

WHAT IT IS
WHERE IT IS
HOW TO KEEP
YOUR HOUSE SAFE

MICHAEL LAFAVORE

 Rodale Press, Emmaus, Pennsylvania

For Trieste, my illumination

Printed in the United States of America on recycled paper containing a high percentage of de-inked fiber.

Senior Editor: Ray Wolf
Editor: Margaret Lydic Balitas
Editorial Assistant: Kerri Balliet
Copy Editor: Linda Harris
Book Designer: Denise Mirabello
Illustrator: Frank Rohrbach

   The presence of radon gas in the home is an important health issue. You can protect yourself only by properly and professionally performed testing and appropriate remedial measures. The publisher takes no responsibility for such activities. The reader is advised to seek competent professional help from qualified experts.

Library of Congress Cataloging-in-Publication Data

Lafavore, Michael.
    Radon : the invisible threat.

    Includes index.
    1. Radon—Toxicology. 2. Radon—Environmental aspects. 3. Air—Pollution, Indoor—Hygienic aspects.
4. Housing and health. I. Title.
RA1247.R33L34   1987   628.5'35   87-4829
ISBN 0-87857-697-5 hardcover
ISBN 0-87857-712-2 paperback

2 4 6 8 10 9 7 5 3 1   hardcover
2 4 6 8 10 9 7 5 3 1   paperback

CONTENTS

ACKNOWLEDGMENTS

Many thanks to senior editor Ray Wolf for believing in this book; John Viehman, executive editor of *Rodale's Practical Homeowner* magazine, for his support and understanding in allowing me time away from my job to write it; Paul Lafavore, my bright kid brother, for helping me fathom the biological effects of radon at the cellular level; the Environmental Protection Agency (EPA) and the Department of Energy (DOE), who haven't always done enough, but who have still produced some first-rate research on radon testing and mitigation, much of which is included in this book; all of the people who gave of their time for interviews, especially Naomi Harley, B. V. Alvarez, Robert Yuhnke, Gene Tucker, Andreas George, Richard Guimond, and dozens of contacts at the EPA; Cheryl Winters Tetreau for assistance with chapter 8; Dhiren Mehta and Robert Flower of Rodale Press for technical support; Margaret Lydic Balitas for her expert editing and kind patience; Linda Harris for her excellent copy editing and close attention to details; Denise Mirabello for her fine book design and layout; Kerri Balliet for her fact checking of names and addresses; and the Kennedys, of Hazleton, Pennsylvania, for popcorn, caring, and a quiet place to work.

1

THE MOST RADIOACTIVE HOUSE IN AMERICA

One way or another, most of us have learned to fear radioactivity. Not all of us understand it completely, but we know that it is seldom good for us and that we should try to avoid it. We've been trained to think of radioactivity as something man-made—a powerful, unstable force, with great potential to do catastrophic harm to us, but a force that is usually containable and controllable.

Then along comes something called *radon*, a "natural" radioactive substance that's not created in some huge laboratory or nuclear reactor, and which doesn't wear fluorescent signs announcing its presence. Sometimes radon sounds like something out of a Grade B science fiction movie: a dangerous, invisible gas that lurks underneath your house and sneaks inside through cracks and crevices in the basement.

The difference is, this isn't some menace from outer space; it's always been among us. And the way to defeat

6

it isn't some magic bullet, but an understanding of what it is and how it behaves. This chapter tells the story of how radon first came to the attention of the American public.

THE ALARMS SOUND

Stanley Watras had little fear of radiation after 11 accident-free years of working as an engineer in nuclear power plants. He had seen a lot, but nothing to shake his confidence that he was safe on the job. So he never paid much attention to the radiation monitor stationed at the exit of the Limerick Nuclear Power Plant in Pottstown, Pennsylvania, where he worked. Until he started setting it off.

The newly installed monitor at the plant looks like a high-tech version of the metal detectors that stall travelers in airports. Only instead of revealing keys and loose change, this machine instantly measures radiation levels on nine parts of the body as workers step through.

The monitor is a necessary stop for all workers leaving the plant, a pause to insure that they have not accidentally picked up an unsafe dose of radiation. Watras, like most of his fellow workers, was accustomed to breezing through the machine on his way to the parking lot.

But one day in early December 1984, he stepped into the machine's detection zone and buzzers began sounding. Assuming a mistake, Watras backed up and tried again. Once more the red lights flashed and the persistent buzzer sounded, so loud it made him wince. A technician approached with a hand-held version of the radiation monitor and ran it up and down Stanley's body. There had been no mistake: The readings showed Watras was highly contaminated with radiation from head to foot.

Within seconds, Watras was hustled off to a nearby decontamination room. His clothes were confiscated and placed in a container designed for toxic wastes. He was handed a stiff brush as he stepped into the shower to try and scrub his body free of the radioactivity. Not completely successful, he spent four more hours sitting alone in the room, waiting for the radioactive particles that covered his skin to decay to acceptable levels. It was plenty of time for him to ponder what had gone wrong. No answers were forthcoming.

Caught up in preparations for the holidays, Watras all but forgot about the incident in a few days. For years he had been lecturing anyone who would listen on the inherent safety of nuclear power plants and how much the country needed them. He wasn't about to let a relatively minor incident shake that belief.

Then one day he tripped the alarm again. More hours alone in the decontamination room. Then, again and again he set off the alarm. "I was blowing off alarms right and left," Watras recalls. "I would always have to call my boss and tell him I was going to be late because the alarm went off."

Just trying to get his job done and get out the door after work every day grew into an ordeal for Watras. Sometimes he'd go into the plant to take care of a 15-minute job and instead spend four or five hours getting rid of the radiation he was mysteriously picking up. On particularly bad days, the monitors showed he was carrying six times more radiation than safety regulations allowed.

But the pieces of the puzzle just would not go together for Watras, no matter how much he tried to make them. For one thing, the Limerick plant was inactive at the time. And even if there *was* an unknown source of radioactivity, why was it affecting only him?

On December 14, 1984, Stanley Watras got his answer. As on every other workday, he got dressed, ate his breakfast, and headed off for the drive to Limerick. But when he got to the plant, he impulsively varied his routine. Instead of heading down the long corridor toward the power block after walking through the front door, he made a 180-degree turn to the exit door and stepped through the radiation monitor.

It went off.

"An alarm was going off in my head at the same time," says Watras. "It was telling me that I wasn't picking the radiation up at the plant. If I was already contaminated when I got to work in the morning, there was only one place I could be getting it. And that place was my house."

But that did not make sense. The house, a modest seven-year-old brick and wood split-level that Watras and his wife, Diane, had bought a year earlier, wasn't even near the plant. Set on a ridge overlooking the rolling countryside in Boyertown, Pennsylvania, it was several miles from Limerick. Watras had never brought anything that could possibly be radioactive home from work, and the previous owner hadn't been connected to the nuclear industry in any way.

Stanley asked his employer, Philadelphia Electric Company (PECO), to send some men over to run tests on the house. Five days later a team of specialists arrived and unpacked their equipment. Almost as soon as one of the technicians turned on his Geiger counter it began clicking. Watras, watching the men work, knew enough about radiation to realize there was trouble.

"Normally that Geiger counter should have clicked about every 20 minutes," he says. "But as I listened, it was going off every 2 minutes. Something was wrong."

The technicians thought so, too. The readings their equipment was showing seemed so preposterous, they did some tests over again. But the results were the same.

The readings in the Watrases' living room showed radiation levels 700 times higher than the *maximum* considered safe for human exposure.

Ironically, the source of radiation in the Watras house had nothing to do with Stanley's work. In fact, the job probably saved the Watrases' lives by alerting them to the problem. The culprit, the researchers concluded, was radon, a naturally occurring radioactive gas given off by underground uranium.

After the technicians left, Stanley and Diane sat at the kitchen table, dazed and frightened by what they'd been told. First came fear for their children: Michael, four, and two-year-old Christopher. Next came questions like "Why us?" Then came the frustration of confronting an invisible enemy that neither of them had heard of before. There wasn't even anyone they could blame for their troubles. "Who were we going to blame?" asks Stanley. "God put it there, and we couldn't very well shake our fists at him."

FIXING A BROKEN HOUSE

The first order of business, they decided, was to find out more about radon and how to stop it from getting into their house. But their attempts at getting answers the next day proved equally frustrating. "No one we called knew much of anything about [radon], or what could be done to fix it," says Diane.

When he saw the reading from the Watras house, Thomas Gerusky, chief of the Pennsylvania Bureau of Radiation Protection, was dumbfounded. "I couldn't believe you could get contaminations like that just from going into a person's house," he says. "I've been in a lot of uranium mines and none of them was that hot."

But skeptical or not, Gerusky sent two men back to the Watras house on January 5. When they pulled into the driveway, Diane went out to meet them. They told her that the house wasn't safe and that the family should find someplace else to stay.

"How soon do you want us to move?" asked Diane.

"Now," she was told.

"How long will it take us to get the house fixed?"

"We don't know if it *can* be fixed," answered one of the men.

It was frigid and snowing outside as the couple took down the Christmas tree and threw some clothes into suitcases, all the while wondering where they could go and if they would ever be able to come back. Neither of them had any relatives in the area, and a family of four couldn't very well just appear on a friend's doorstep spinning wild tales of a radioactive house and men in government cars.

So, late that evening they checked into a local Holiday Inn and put the kids to bed. In the morning there wouldn't even be any Christmas presents for the children to play with; the toys would have to be decontaminated first. The long nightmare was beginning.

"Our world just began collapsing in front of us," says Stanley. "It was really a tormenting time. We were dealing with the unknown and it was horrifying."

Some of what they were able to learn just frightened them more, such as the estimate from the United States Environmental Protection Agency (EPA) that living in the house had posed about the same health risks as smoking 280 packs of cigarettes a day, or that the single year the family spent in the house increased their chances of getting lung cancer by 13 or 14 percent, or that if they'd lived there for eight or ten years without discovering the radon they all would have certainly contracted lung cancer. (The previous owner, who *had* lived there for several years, has so far declined to talk publicly about the house.)

"I don't smoke and I never touch alcohol when I'm pregnant," says Diane. "We've tried to be careful, but then we brought our newborn baby to live in a radioactive cloud. And no one knows for sure what it might have done to us."

After five weeks at the motel, and still with no idea whether they would ever be able to live in their house again, the Watrases reluctantly rented an apartment. The financial burden was enormous; on one income, they would have to buy food and essentials, pay rent on the apartment, *and* keep up mortgage payments on the house. They'd already talked to the bank and their insurance agent, but neither had ever heard of radon. They wouldn't be able to help.

The winter slipped by without a resolution to their problem. Stanley stopped by the house occasionally to pick up something, but he wouldn't let any of the family inside, even for a minute. "From the moment we left until the house was fixed, my children didn't spend another second there," says Stanley. When the kids asked why they couldn't go home, Diane told them the house was "broken and needed to be fixed."

Finally, on April 5, 1985, PECO announced that it would do the fixing by funding a cleanup operation on the house. In the process, it would become an on-site laboratory to study radon.

And there was plenty to study. No one had ever seen a radon problem this bad, nor even imagined it was possible. No one was sure what would work; trial and error was the only way to approach the project.

Researchers moved their sophisticated equipment in and began 24-hour monitoring of radon levels all over the house. They arrived at a theory that radon gas from the soil around the foundation was dissolving in water and seeping through the basement walls. A contractor was hired to try and fix the problem by excavating earth from around the concrete walls and installing plastic drain pipes to carry away groundwater before it could get into the house. The strategy lowered radon levels by about 15 percent, which wasn't nearly enough.

Next, the foundation walls and floor were coated with a special epoxy liquid designed to seal any cracks in the concrete. The sump pump hole in the basement was also sealed. Then a special ventilation system was installed, which caused another reduction in radiation levels, but the radon levels remained dangerously high.

Finally, the researchers tried a last-ditch, and expensive, effort. The plan was to support the house on jacks and dig up the entire foundation, then lay down a plastic barrier that would block radon gas, install another ventilation system, and build a new basement. Then workers donned protective masks and clothing and began breaking up the foundation.

Stanley and Diane Watras pose with their two children in the front yard of their house in Boyertown, Pennsylvania. When they bought it, the house was so contaminated with radioactive radon gas that living in it was as dangerous as smoking 280 packs of cigarettes a day. The family still lives there, but the radon is nearly all gone. (Photo by Mitchell T. Mandel.)

Soon after breaking through the concrete, their jack-hammers hit solid bedrock. More tests solved the riddle of the most radioactive house in America. When the builder had hit the same rock seven years before, he had decided it would make a solid foundation for the house and had poured the concrete basement floor right over it. He had no way of knowing that the glittering yellow "rock" was actually a 40-foot-wide vein of uranium ore.

"They had low-grade uranium ore beneath the basement of the house, in direct contact with the house," the manager of the company doing the remedial work told a local newspaper. "Obviously that's where the radon was coming from." Sandwiched on its sides by two layers of impervious rock, the uranium was continually pumping enormous amounts of radon gas directly into the basement, and from there the gas quickly spread to the rest of the house. Had the house been built only 4 feet to the left, the radon would have escaped harmlessly to the outdoor air and Stanley would never have set off radiation alarms.

Since moving the house wasn't a viable option, the workers continued to remove the foundation, then went further and dug up soil and rock from 4 feet below. (The refuse was classified as a toxic waste and had to be specially disposed of.) Then they installed a new ventilation system of fans and stacks that ran from under the new foundation to the very peaks of the home's roof. The fans were set to run constantly, to pull radon-laden air out from underneath the house and exhaust it through the vent stacks. Diluted by the outdoor air, the radon gas is rendered harmless.

Another round of tests showed that, finally, the house was safe enough to live in. The cleanup had taken months and cost PECO $32,000, but they'd learned a lot about how radon gets into a house and how to get it out.

Six months after they were told to leave, the Watras family moved back into their house. They're still there, assured by scientists and health experts that they're not in any danger. But no one is taking any chances. The Watras family lives among portable radon detectors, one in almost every room. They pray that the radiation levels won't jump again, like they did a few months after the repairs were completed, when one of the ventilation fans wasn't working properly.

The house is their home—probably for a long time. They never consider the possibility of selling the house, for which they paid $67,500 and to which they added thousands in improvements. After months of television newspersons camped out on the lawn, photographing strangely garbed workmen, the house has become notorious as a local tourist attraction.

"We think it's worth absolutely nothing," Stanley told the *Philadelphia Inquirer*. "But even if we could sell it, I don't think we could morally sell it."

THE STORY CONTINUES

What happened to the Watras family is a sad and frightening story. But if the house were a mere oddity and the Watrases simply a hapless couple who picked the wrong place to buy a house, the story could end right here.

It's not nearly as simple as that. The Watras house is far from an isolated case. Radon is a national issue. The official EPA estimate is that 8 million homes in the United States may contain unsafe levels of radon. Other experts fear the number is much higher. One EPA official, who asked not to be identified because his views differ from the agency's official estimates, says he wouldn't be surprised if 10 million homes were affected.

After news about the Watras house sent shock waves through homeowners in the area, more than 20,000 houses in Pennsylvania alone were tested for radon; nearly 60 percent of them showed unsafe levels of the gas. The shock waves spread beyond Pennsylvania. In Florida, the first radon law is already on the books, requiring home builders to meet tough regulations designed to keep radon out of new houses.

Radon could be responsible for as few as 5,000 and as many as 30,000 lung cancer deaths yearly in this country. William Belanger of the EPA says it's "our worst environmental threat," ten times more dangerous than asbestos or toxic wastes. No home is immune to the possibility of high radon levels. Every state has areas of radon contamination that could pose a health threat, and trouble spots have been identified in Canada, Europe, and other parts of the world.

Stanley Watras discovered the trouble with his house by accident. Fate put him in a terribly radioactive house and at the same time in a job that would warn him of the danger. Now he says, "I pray no one else has to go through what we have."

But that's not likely. "It's unrealistic to hope that the first really bad house we find will be the worst case," says Richard Guimond, of the EPA. "There are probably others out there that are as bad . . . or worse." These other houses don't have owners who work in nuclear power plants, or whose employers are willing to pay for repairs. What about these homeowners? How can they find out if their homes are contaminated with radon? And what can they do if there is a problem?

We are in luck, though. As United States Senator Frank Lautenberg (a Democrat from New Jersey) points out, "Radon is a treatable hazard." Although there are still many unanswered questions, thousands of man-hours and millions of dollars have gone into the study of radon during the past few years, many of them since the Watras house was discovered to have high levels of radon. We now know a great deal about the best ways to test homes for radon and how to best fix the ones with problems. That's what this book is all about.

CHAPTER

2

WHAT IS RADON?

Understanding this invisible threat known as radon means finding out what it is, where it comes from, and how it finds its way into houses. The best defense against it is a clear understanding of why it behaves the way it does.

It's fascinating that we've taken so long to "discover" this natural form of radiation, although it's always been with us, and will be forever. To read the headlines of many newspapers these days, you might get a different idea and think that radon is something that just recently appeared out of nowhere to menace homeowners.

The story, however, begins a bit further back in time—about 4.5 billion years earlier, actually. That's about how long there's been an earth, and as long as this planet has been around, so has radon. It's just that people didn't give radon a whole lot of thought until they had to.

Let's start at the beginning, the *very* beginning. Don't worry, this is the short version; this entire book wouldn't have room enough to explain in detail the theories on how the earth was formed, and they're probably not all that important to you and me anyway.

The subject is important to scientists, though. They like to spend a lot of time at conferences and conventions arguing their theories. One very popular one goes like this: Billions of years ago, a catastrophic explosion on a huge star somewhere ejected an enormous cloud, made up of 99 percent gas and 1 percent dust, into space. The planets in our solar system, the story goes, somehow condensed and took form out of this rotating disk.

When the dust of creation had finally settled, the earth ended up as a spherical planet with three layers: a core of solid and molten iron, a 1,800-mile-thick intermediate mantle, and an outer crust of rock, as little as 5 miles thick in some places and as much as 25 miles thick in others.

This crust is made up of a jumble of different elements, mainly oxygen, silicon, aluminum, iron, magnesium, and calcium. These are all relatively stable and dependable; you can count on them to be the same tomorrow as they were today. A given atom of oxygen would have looked and acted the same a million years ago as it does today.

URANIUM

Way down on the list of the planet's ingredients, making up a tiny 4 parts per million (ppm) of the crust, is a very heavy and complex element that has the distinction of being the great-great-grandfather of radon. It's called uranium-238.

First discovered in 1789 and named, for no particular reason, after the planet Uranus, uranium is anything but stable. Since the beginning of time, it's been in a constant state of change, continually throwing off energy, breaking down and transforming itself into one simpler substance, then another.

Put another way, uranium is radioactive. That means that as the unstable atoms of the element break down into atoms of more stable elements, a process known as decay, they spontaneously give off bits of pure energy, known as radiation. There are three types of radiation related to the decay of uranium: alpha particles, beta particles, and gamma rays.

ALPHA PARTICLES

These consist of a nucleus of two protons and two neutrons and have a positive charge. Because of their large size, alpha particles have a range in air of about 3 inches and little penetrating power; they can be stopped by a piece of cellophane or the first layer of skin on the outside of your body. Think of an alpha particle as a cannonball: It's big and doesn't travel far, but it can do a lot of harm. Alpha radiation disperses its energy quickly, damaging the molecules as it passes through.

BETA PARTICLES

If an alpha particle is like a cannonball, a beta particle is like a rifle bullet. Emitted from the nucleus of a radioactive atom, it moves at nearly the speed of light and can travel several feet in the air. Beta particles have moderate penetrating power; they can pass through ¼ inch of Plexiglas, and slightly less than that into surface tissue.

GAMMA RAYS

Gamma radiation is the most energetic type. Unlike alpha and beta radiation, which are particles, gamma radiation is an electromagnetic parcel of photonic energy, like a ray of light. Similar to an X ray, this energy pulsates and oscillates like a wave. Gamma rays have great penetrating power; they can saturate the human body and even pass right through it.

Exposure to all three types of radiation is dangerous to living tissue. But because alpha radiation moves slowly and is denser, its biological damage per unit of energy is many times greater than that of beta or gamma radiation.

THE DECAY CHAIN

The time it takes uranium, and every other radioactive element, to break down and release its radiation is measured as its *half-life*: the time required for half of its atoms to decay. Let's say an element has a half-life of one year. If you put a pound of it into a box and opened the box only after a year, half a pound of the element would be gone. If you resealed the box and waited another year, you'd be down to a quarter pound. At the end of the third year, an eighth of a pound would be left, and so on. The element would never completely disappear, but what remained would be reduced by half each year.

Uranium-238 has a half-life of 4.5 *billion* years. So, whereas it's true that all the uranium that will ever exist already does, it will always be with us. In the first link of its long decay chain, uranium shoots off two protons and two neutrons and produces a "daughter" element, thorium-234, with a half-life of 24 days. Then, thorium expels a beta particle and transforms itself into something called protactinium-234, with a half-life of only 1 minute.

About halfway through the decay sequence, uranium sheds another identity and becomes radium-226 (half-life of 1,602 years), the stuff that was used to make watch dials glow before watchmakers stopped using it and switched to something a little less radioactive.

It's at the point where radium itself begins to decay that the long decay chain of uranium begins to cause some real problems for people. This is because radium, as it disintegrates, breaks the chain of solid elements and turns itself into a gas, called radon.

RADON GAS

First discovered in 1900 by German physicist Friedrich Ernst Dorn, this highly radioactive gas was originally called simply, "radium emanation." Dorn thought that radon, as it was later named, was very rare and chemically inactive, so he believed he had found the sixth noble gas (The others are helium, neon, argon, krypton, and xenon).

He was partly right. Although we now know that it's far from rare, radon does resemble other noble gases in that it's colorless, odorless, tasteless, and nonflammable. And radon is chemically inert. It doesn't react or combine with other chemicals or elements, nor is it readily absorbed by them. It can travel long distances through rock, soil, and water, and emerge unchanged.

As a matter of fact, although people say they're very concerned about radon today, radon gas itself is relatively harmless. Because it's inert, and a gas, you breathe it in and you breathe it out again—it doesn't stick. Radon, though, has some very dangerous daughters.

RADON DAUGHTERS

The decay products of radon, all but one of which have half-lives of less than 30 minutes, are all radioactive particles. As radon releases its ionizing radiation, it's transformed first into polonium-218, which then decays into lead-214, followed by bismuth-214, polonium-214, and lead-210. A final decay, and the end of this chain, is the change of lead-210, which has a half-life of 22 years, into a stable, nonradioactive molecule, lead-206. (The number after an element is its atomic number. It refers to the combined total of protons and neutrons in the atom. Even when two elements have the same name, if their atomic numbers are different, they are different elements.)

Lead, bismuth, and polonium are heavy metals, and they're far from inert. They're chemically active and behave like ultrafine particles in the air. And because they're electrostatically charged, they quickly attach themselves to dust and smoke in the air, clothing, furniture, and walls. But what makes these radon daughters so dangerous is that when they're inhaled, some of them attach themselves to air passageways in the lungs and upper respiratory tract, where they continue to decay and shoot off bursts of alpha, beta, and gamma radiation.

URANIUM IN ROCKS AND SOIL

Wherever you find uranium in rocks and soil, you'll find radium. And where there's radium, radon gas and its daughters are sure to be found. There's some uranium in almost all rocks and soil in America, but the amounts are usually small. The United States Environmental Protection Agency (EPA) estimates that the average soil here contains only about 1 part per million of uranium.

But averages are arrived at by factoring together highs and lows, so they don't tell the whole story. Some parts of the country have far more uranium in their soil, and much more serious problems with radon gas, than

Radium Decay Chart

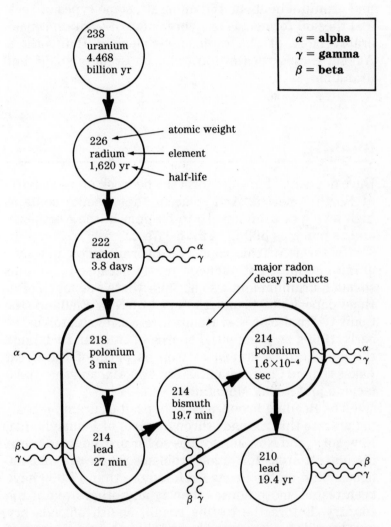

Shown here is the decay chain that transforms uranium into radon and its daughters. About halfway through its decay sequence, uranium becomes a gas, radon, which as it disintegrates gives off radioactive particles of polonium, bismuth, and lead. Also noted is the half-life of each material—the time it takes for half of its radioactivity to dissipate.

averages indicate. The reason is that, although you can find uranium in about 150 minerals, some types of rock, and the soil formed as they break down, contain considerably more of the radioactive element than others. Among these types are granite, phosphate, shale, and uranium.

GRANITE

Huge deposits of granite form the bedrock of many parts of North America. And some of that granite contains high levels of uranium, 10 to 50 ppm in the Northeast, and as much as 500 ppm in the West.

In terms of this country's overall radon problem, granite is the worst offender, partly because it contains so much uranium and also because there's so much of it. Huge deposits of it underlie most of New England (the Conway granite of New Hampshire was evaluated in the early 1960s as a potential source of nuclear fuels) and the mid-Atlantic states, as well as parts of the Great Lakes region, California, Eastern Canada, and the Appalachian and Rocky Mountains.

The Reading Prong, a large deposit of uranium-bearing granite that extends through parts of Pennsylvania, New York, and New Jersey, has so far proved to be one of the hottest areas for radon problems. There are 320,000 homes built on the Prong. So far, more than 17,000 have been tested, more houses than in any other area of the country. Before the testing began, health officials privately said they expected to find that about a quarter of those houses contained potentially dangerous amounts of radon gas. They were wrong. When all the tests were tallied, nearly *60 percent* of the homes had exceeded the safety limit.

PHOSPHATE

Phosphate rock is a source of radon similar to uranium. Deposits of phosphate often contain high levels of uranium—50 to 150 ppm. That's unfortunate, because the rock is the main source of phosphorus, one of three essential ingredients in chemical fertilizers, and is used to make building materials, such as phospho-gypsum. One-third of the world's phosphate comes from north and west-central Florida, but there are also mines in Idaho, Montana, North Carolina, Tennessee, and Wyoming.

Not only is the phosphate rock itself a problem in Florida, but its mining and milling may also release significant amounts of radon into the environment. The amount of uranium contained in the annual production of processed phosphate rock is about equal to that mined as uranium ore. Usually strip-mined, the phosphate rock is then crushed and screened to separate it from the attached clay and sand, which are often returned to the mine area as fill. The land is then shaped and contoured to make it look like the bulldozers and dump trucks never had been there.

But there are unseen changes. The mining process brings once deeply buried radioactive rock closer to the surface, and reclamation spreads the leftovers, which still contain 10 to 50 percent of the original uranium, across a much wider area than they originally covered. So it's not surprising that reclaimed phosphate soil has shown radon levels seven times higher than undisturbed phosphate land.

What is surprising is how much of that reclaimed land is somebody's backyard. As population pressures have squeezed Florida, reclaimed phosphate land has been turned into housing developments, shopping centers, and industrial parks. Despite an EPA warning that "Structures built on reclaimed [phosphate mining] land have radon daughters significantly greater than those not built on reclaimed land," more than one-third of Florida's 25,000 acres of reclaimed land have been built on. In 1986, a state law took effect requiring minimum ventilation rates in homes in these high-risk areas.

A 1981 comparison of cancer rates in the 48 contiguous states by researchers from General Electric found that areas with large phosphate deposits have a higher incidence of lung cancer than areas where there's little or no phosphate.

SHALE

Shale is nothing more than hardened and compacted clay and mud, so its uranium content depends on the soil it was created from, which varies dramatically from place to place. In Sweden, where you're not allowed to build a house until the lot has been tested for radon, shale is the main culprit. In this country, some deposits, among them the Chattanooga Shale in Tennessee, are high in uranium.

URANIUM

Large deposits of uranium ore, pure enough to mine for atomic fuel, are located in parts of western Colorado, eastern Utah, northeastern Arizona, northwestern New Mexico, Wyoming, Texas, and western Canada. And there are probably isolated deposits, such as the one found under the home of Stanley Watras, in many parts of the country.

Irregular in shape and size, the uranium deposits range from small masses, no more than a few meters wide and thick, to chunks of ore hundreds of meters wide and thousands of meters long.

As you might expect, radon levels in the soil near these uranium deposits are high. The ore contains about

1,000 times more uranium than the average soil. In fact, uranium prospectors measure radon levels in soils to help them locate veins of the mineral.

NUCLEAR WASTES

In addition to problems from the ore itself, there's also concern about the processing of wastes left over after uranium is milled into fuel for nuclear power plants. The tailings, which still contain as much as 15 percent of the uranium in the ore, are often discharged from the mill as slurry that is stored in large piles. Currently, those piles are growing at a rate of 25 million tons a year. Heavy storms can cause the piles to collapse and contaminate rivers and streams; rains leach out radioactivity into groundwater. And the piles will continue to give off radioactivity for hundreds of thousands of years.

Some studies have shown that more radon gas is released from uranium waste dumps than from deposits of uranium ore. Measurements of the radioactivity emanating from tailings have shown just how hot things can get. In Colorado, one pile was found to be pumping out enough radiation to raise levels in the air around it to 2,500 times higher than normal.

That's frightening enough if you live anywhere near a tailings pile. But what if you lived right on top of one? Thousands of people do. As astounding as it now seems, tailings were once used by unsuspecting builders and homeowners to make concrete and building materials, under foundations of houses as backfill, and to reclaim land in the western United States and Canada.

If you could see a pile of tailings, you'd understand why. The stuff looks great to builders and contractors. It's the kind of thing a busy contractor could find a lot of uses for. The uranium companies, on the other hand, saw it as a nuisance that just got in the way. They were more than glad to give it away to anyone willing to haul it off. And haul it people did, by the truckload.

As much as *10 million tons* of the tailings were used in one city alone, Grand Junction, Colorado. In their book, *The Menace of Atomic Energy* (see the bibliography), Ralph Nader and John Abbotts say that between 1952 and 1966, hundreds of houses and businesses were built with radioactive tailings from the Climax Uranium Company in Grand Junction. Apparently, the tailings were used mostly as construction fill underneath or against the buildings, "although in one instance—a school—the masonry itself was made with tailings."

A survey of the Grand Junction area later indicated that thousands of buildings were affected. And not surprisingly, lawsuits began to fly when the problem was finally uncovered decades after the tailings were used. The government and the mining companies have agreed to fix the worst houses in the uranium-mining towns, but the case is still clouded by controversy.

The frightening part is that many of the people who lived there, or who still live there, don't know for sure what the radiation might have done to them. All they have to go on are theories like the one expressed by Dr. John W. Gofman in his book *Radiation and Human Health* (see the bibliography). He says, "It would be surprising if a serious lung cancer risk did *not* accompany living in residences built with materials containing high levels of radium. Clearly, the risk is not trivial."

Newsweek magazine estimates that it may cost the United States government $10 billion to clean up the mess at Grand Junction.

Meanwhile, the piles of tailings continue to grow, and, perhaps, to spread radioactivity. According to *Consumer's Research* magazine, scientists from the EPA have estimated that the mill waste generated in fueling one large nuclear power plant could eventually kill about 200 people with its radon emissions.

AREAS WITH POTENTIALLY HIGH RADON LEVELS

The United States Department of Energy is sponsoring work at the Lawrence Berkeley Laboratory (LBL) in California aimed at developing a map of the United States showing potential radon problem areas. That map will certainly be welcome when it comes, but the research probably won't be complete until sometime in the late 1980s.

Meanwhile, the best we have to go on is the map on page 30, compiled by the EPA from geological data. The shaded areas are known deposits of granite, phosphate, shale, and uranium—parts of the country that are *potentially* at risk from radon.

Scientists, and the EPA itself, really aren't comfortable with this map. They published it because people wanted them to—it's the only thing right now that even offers a *clue* to where radon might be found. But at the same time it released the map, the EPA issued a disclaimer for it, which reads, in part, as follows:

> This map should not be used as a sole source for any radon predictions. This map cannot be used to predict locations of high radon in specific locations or individual homes with high radon levels.
>
> Local variations, including soil permeability and housing characteristics, will strongly affect indoor radon levels.
>
> This map is only preliminary and will be modified as research progresses.

So there you have it; it's a map, but no one wants to say exactly what it's a map of. At this point, there's not enough information available to say for sure that most of the houses within the darkened areas will have radon problems, or that areas outside of those shown won't.

Radon testing of houses so far has uncovered areas of contamination in just about every state, and hot spots of contamination in Arizona, Arkansas, California, Colorado, Idaho, Illinois, Maine, Maryland, Massachusetts,

Areas with Potentially High Radon Levels

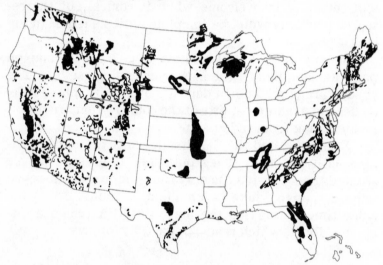

This map, released by the United States Environmental Protection Agency (EPA), shows deposits of granite, uranium, shale, and phosphate—all potential sources of radon gas. Although released under the title "Areas with Potentially High Radon Levels," the EPA takes pains to caution that it "should not be used as the sole source for any radon predictions." If you live in one of the darkened areas, you won't necessarily have a radon problem in your house, and if you're not in a shaded zone, you could still have high levels of radon in your house from a localized source. Other factors, such as the way a house is constructed and the porosity of soil underneath it, can also have a great impact on how much radon will get into the house. (Source: United States Environmental Protection Agency.)

Montana, New Hampshire, New Jersey, New Mexico, New York, North Dakota, Ohio, Pennsylvania, Rhode Island, Tennessee, Vermont, Virginia, Wyoming, and the Pacific Northwest, along with Western and Eastern Canada and Nova Scotia. One study even found 21 "hot" properties in midtown Manhattan.

New areas of radon contamination are being discovered monthly, as more and more homes are being tested. More than 100,000 houses have been tested so far, but considering that there are more than 70 million homes in the United States, we haven't even reached the point yet where educated guesses are being made, at least not publicly, about exactly where people should be worried and where they're safe from radon's threat. Reliable information may not begin to become available until sometime in the 1990s.

But if geologists know where uranium-bearing rock is located, and that rock of that type is continually giving off radon gas, why can't they predict, to the town, to the street, and to the house, where radon contaminations will be found? Because radon defies easy predictions. For instance, although the Watras house showed radon levels 2,153 times higher than normal, the house next door, not 100 feet away, was tested and found to be free of radon. But across the street and 500 yards down the road, another house turned up dangerously radioactive. "No one has ever successfully predicted high radon levels in a particular area," says Dr. Bernard Cohen, a physicist and radon expert from the University of Pittsburgh. "And nearly every problem area has been discovered by accident."

If you don't want your house to be an accident waiting to happen, there's at least one way to find out for sure how much radon you are living with: Test for its presence in your home yourself. As you'll find out later in this book, there are a number of ways to test a house for radon, some of them costing less than $20.

HOW RADON TRAVELS

There are dozens of environmental factors that determine where radon travels, and how far, after it leaves its geological source. But generally speaking, most of the radon gas that reaches the surface originated only a few feet below.

Whether the source is solid rock or, more likely, soil that contains bits of radioactive rock, radon's primary method of transport is simple diffusion. As the radium decays and forms radon, the gas fills the minute air spaces between grains of soil and moves in all directions. But it can travel only as far as its 3.8 day half-life will allow, which is usually no more than a few yards. If the source is closer than that to the surface, the gas escapes to the outdoor air.

Underground caves, cracks, fissures, and faults in the earth can act as natural pipelines, funneling large amounts of radon more quickly to the surface from deep below. There's evidence that radon can migrate to the outside from at least 170 meters underground with its radioactivity intact.

There are loads of environmental factors that also help determine how much radon gets to the surface. A difference in temperatures between the air and ground causes a pull as warm air rises, creating a slight vacuum behind it that can draw radon out of the soil. The same effect takes place on windy days, when air passing over the ground surface creates a suction in the top layer of soil. Radon emanation is usually highest near sunrise and at midafternoon, when the atmosphere is most turbulent.

Several tests have shown that falling barometric pressure tends to pull more radon out of the ground. One study showed a doubling of the radon emanation from the soil after a mere 1 percent drop in barometric pressure.

Some researchers say the permeability of the soil may be one of the most important factors of all. Loose or sandy soils allow radon gas free range because of the large number of air pores in them. Clay and other highly compacted soils, on the other hand, are less likely to transport radon as easily, because there is less air in them for radon to move through. Similarly, dense rock near the surface, provided it doesn't contain uranium or large cracks, will stop radon better than porous stone. In two monitoring studies (one in New Jersey and the other in Tennessee), houses in the same area were measured for indoor radon levels. In both tests, 30 to 50 percent of the homes built on porous sandstone or dolomite deposits had unsafe radon counts, while nearby houses built on more solid rock were normal.

Some studies conducted at LBL indicate that if soil permeability is high, houses may end up full of radon, *even if the radium content of the soil is moderate to low.* "If you look at the degree to which different factors contribute to houses having high concentrations of radon, you'll find that the permeability is the single most important, and most variable factor," says Anthony Nero, of LBL. "The radium content of the soils in different parts of this country might vary by a factor of 10 from place to place. But the differences in the permeability of the soil from one house to another can easily vary by a factor of 1 million."

Weather also affects how much radon can get through soils. When it rains, moisture fills the soil pores and cuts the movement of the gas by two-thirds. The same thing happens when the ground freezes. Measurements usually show lower levels coming from the ground during the winter, followed by a sharp jump after the spring thaw.

Then again, there's the "snow effect" to contend with. A radon sleuth in Colorado discovered that snow piled up against the house can increase indoor radon levels, and shoveling the snow away can decrease them. The theory is that the snow traps radon as it comes out of the ground, blocking its path to the atmosphere. At that point, the easiest place for the radon to go is into the basement.

RADON IN WATER

Another wild card is radon's sneaky habit of traveling long distances by hitchhiking a ride in water. Radon gas is fairly soluble in water. As groundwater travels through soil and rocks that contain uranium and radium, it picks up radon gas along the way. How much of the radioactive gas dissolves into the water depends on the source. But if the water is passing through radium-containing deposits of granite or other radioactive minerals, as it does in Arkansas, California, Florida, Georgia, Massachusetts, North Carolina, Oklahoma, Pennsylvania, Texas, Utah, Virginia, and much of the eastern United States and Canada, you can practically bet on elevated radioactivity.

There is no need to worry very much about radon in water that comes from large municipal water systems. This is because radon daughters are short-lived; most of the radioactivity will have dissipated in the hold-up and treatment time before the water reaches the tap.

Smaller systems—those serving fewer than 1,000 people—can be a different story. Holding times are shorter, so less radon decays away before the water reaches homes. The EPA says as many as 30,000 of these public systems may be delivering high doses of radon along with their water.

Wells are the most direct and immediate link between groundwater and houses, and are the sources of the worst radon contaminations from water. More than a quarter of all wells in the United States have radon levels twice as high as the water in the average municipal system. In Maine, New Hampshire, and Rhode Island (states with huge granite beds), *most* wells have elevated radon levels. When 2,000 Maine wells were tested, the average radon concentrations were found to be 5 times more than the national average. Using water samples from wells in seven Maine counties, another survey found that 98 percent of them exceeded the EPA safety limits for radon in water. In some wells in that state, radon levels 50 times higher than normal have been recorded, and in one the radon level was more than 100,000 times higher.

"The more we look, the more we find," says Jerry Lowry, of the University of Maine Department of Civil Engineering, who believes that all the Eastern states as far south as North Carolina could have high levels of radon in groundwater.

RADON IN NATURAL GAS

Natural gas (methane) will also pick up radon as it travels through the earth, although the problem isn't causing as much concern as radon in water. Radon levels in natural gas vary, according to where it's produced. The Gulf Coast region, where 40 to 60 percent of the country's natural gas comes from, has the smallest amount of radon in its gas. In Colorado, Kansas, and New Mexico, the amount of radon in the gas may be 5 to 20 times higher than it is in Louisiana and Texas. According to H. Ward Alter, president of Terradex Corporation, one of the largest radon testing firms, problems with indoor radon that may be coming from natural gas have shown up in houses in Ohio.

The radon in natural gas is further neutralized as it is stored, pumped through trunk lines, and mixed with gas from other wells. Much of the radioactivity has decayed by the time the gas reaches consumers, unless, of course, they live close to the wellhead.

All natural gas undergoes some processing before it's distributed, and that usually lowers the radon concentrations. The exception is liquid petroleum gas (LPG). During the processing of natural gas, impurities are removed and hydrocarbons are separated out. Some of these hydrocarbons get bottled and sold as LPG. And most of the radon goes along with it, because it tends to leave with the hydrocarbons. Tests in New Mexico showed that radon levels in LPG were 20 times higher than in the natural gas it was made from.

IS IT SAFE TO BREATHE?

With all this radioactive gas leaking out of the ground in so many ways, you'd think that it would be dangerous to take a deep breath outdoors. But the numbers are on our side. There's a lot of air out there to dilute the radon. As a result, there are only about 0.00000000000006 atoms of radon in every million molecules of air. This background radiation is something we're all exposed to every minute of every day, but its impact on our health is relatively small.

Scientists say that radon gas entering the outdoor air from underground increases the average person's lung cancer risk by about one chance in a million for each year of exposure.

But the odds, and the risks, can change dramatically when a house gets in radon's way as the gas comes out of the ground. That's when radon stops becoming some obscure geological phenomenon and starts becoming something millions of people need to be concerned about. Because if there's one fact that everyone is sure about, it's that houses can bottle up radon, allowing it to build up to dangerous levels.

CHAPTER

3

RADON
IN OUR HOUSES

Millions of houses in the United States and elsewhere are sitting ducks for radon. Some are leaky old antiques, others are new homes built as tight as ships. They are ranch houses, underground houses, solar houses, condominiums—all types of structures. They only have one thing in common: They're in the wrong places.

What's the wrong place? Wherever the uranium content of rocks or soil is higher than average. That's because if there's radon being produced underneath and around a house, it's probably going to find a way to get indoors. Given the minutest point of entry, the gas can seep into a house at an alarming rate. And where the air outdoors tends to dilute radon and render it harmless, a building can bottle up and concentrate it, making it much more dangerous.

If you upend a bucket over a piece of ground that contains radon gas, in a short while the radioactivity levels inside the pail will be much higher than in the air outside it. Even if you punched a few holes in the top, radon would be entering at the bottom more quickly than it could escape through the top, and radon levels would remain high.

Many houses are like that bucket. Built over a source of radon, they fill up with unnaturally large amounts of the gas. New radon enters before what's already there can escape outdoors through windows, doors, and cracks in the building shell. Trapped inside, the gas continues to concentrate, decaying to radioactive particles that can make breathing the air in the house an unhealthy proposition.

Researchers say the amount of radon in most houses is ten times higher than outdoors. And, as witnessed by the Watras house and others tested around the world, levels can reach hundreds, even thousands, of times higher.

Since most soils contain some radium, there's at least some radon in nearly every house in this country. But how does it get inside? And why are some houses relatively safe, and others seriously contaminated?

No simple answer will work. There are a number of factors that determine how radon finds its way into a house, how much will get in, and how long it will stay there. It's important to understand them all, because radon doesn't ring any bells to announce its presence. This is an enemy you've got to know well to defeat.

Although water, natural gas, and some building materials can contribute to indoor radon problems, by far the most significant source of trouble is radon in the soil and rocks under and around the house. The gas seeps in through cracks and holes in the foundation.

DIFFUSION AND FLOW

As radium exhales radon, it becomes part of the gas that fills capillaries between soil particles. After that, it can travel through the ground in two ways: *diffusion* and *flow.*

Simple diffusion usually doesn't get radon very far. The gas moves from space to space between soil particles, traveling no more than its diffusion length, which is the average distance an atom can move through dry soil before it decays and becomes harmless. Usually that distance is no more than a few feet in normal soil, more in porous soils with lots of sand in them.

Diffusion is a very inefficient method of travel, and if that were the only way radon moved through the soil, few of us would have anything to fear from it. But when other factors are at work, radon travels much further and more aggressively.

Pressure-driven flow, instigated by wind, temperature, and barometric pressure, can push radon gas through the soil and into houses, over longer distances and with greater force. Decreasing barometric pressure causes an upward flow of air through the soil and increases radon levels in the top layers of soil. In one study, a moderate drop in barometric pressure caused radon levels in a house to rise to a concentration five times higher than normal. (A rising barometer can have the opposite effect, bringing about a draw-down of air into the soil and pushing radon deeper into the ground.)

Flow can also be intensified by the *stack effect* (as in smokestack), a convective current caused by temperature differences in the house, especially those that occur during the heating season. Inside the house, warm air rises, eventually flowing out of the attic and the upper stories through unsealed windows, the roof and chimney, and any cracks and crevices it can find in the building shell. The rising air current creates a slight vacuum in the basement and lower parts of the house. In turn, that lower pressure allows radon gas, at a higher pressure in the soil under and around the house, to push into the basement through even the tiniest of openings. Thus, like air being drawn up the chimney of a lighted fireplace, the stack effect pulls radon-bearing air from the ground.

Indoor radon concentrations can vary greatly between summer and winter. University of Pittsburgh physicist Dr. Bernard Cohen, who has collected results from testing the radon levels in more than 32,000 homes, says winter radon levels are about 60 percent higher than summer levels.

Finally, wind can drive radon from the soil. A good breeze raises the air exchange rate between the house and the ground and increases the pressure differential across the building shell. The differential, known as

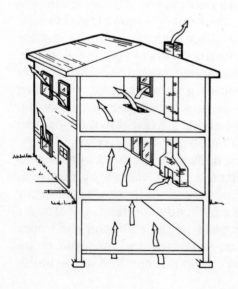

"The stack effect" is a phenomenon caused by warm air rising through the house and escaping out cracks and windows in the upper stories and attic. The air current caused by this movement creates a negative pressure in the lower sections of the house, which in turn can draw more radon gas into the basement.

wind load, causes air to flow in from the soil on the windward side of the house, and out into the soil on the opposite side, carrying radon along with it.

Once the radioactive gas reaches the space underneath a house, it usually doesn't have much trouble getting inside, particularly if it's worked up some speed from a pressure-driven flow. All it takes is one tiny crack, so small it may be invisible, and you can have a houseful of radon.

THE INTRUDER BELOW

Nearly all houses in North America are built on one of three types of foundations: a concrete slab poured on grade; a wooden floor over a crawl space; or a basement.

If there's a strong source of radon under the house, slab-on-grade construction offers the most protection from the gas. A 4-inch-thick slab, provided it's in reasonably good shape and has few cracks and penetrations, blocks much of the radon underneath a house from entering the structure.

Houses with crawl spaces are somewhat more susceptible to radon, especially if the space has few vents to keep air moving in it. Although crawl spaces "disconnect" the house somewhat from the soil below, radon can still be drawn into the house via the stack effect. The less ventilation in the crawl space and the more openings in the floor above it, the higher the amount of radon that will get into the living space.

Basements do the poorest job of keeping radon out of houses. They put such a large surface area of the house into direct contact with the soil that radon can't help but find its way in *somewhere.* The more cracks, crevices,

pipe penetrations, and openings to the soil that a basement has, the greater the potential for radon contamination.

Dr. Cohen's test results show that "houses with
basements have somewhat higher radon levels" than any
other type of construction.

Many houses with basements have underground
foundation walls built from masonry blocks. It's one of
the most popular methods of construction used today.
The rectangular concrete blocks are lightweight and inexpensive, easy to work with and fairly durable. They
also provide the perfect pathway for radon gas to get
into a house—directly up the space formed by the inter-
connected hollow centers of a wall of blocks.

It happens like this: Radon rising up out of the
ground hits the concrete slab that forms the floor of the
basement. This is definitely not a path of least resistance, since the slab is 4 inches of reasonably solid concrete.

So instead, the radon escapes out from under the
slab by flowing through cracks, joints between blocks, or
through the tiny pores of the block itself. Once inside the
dark, spacious "chimney" formed by the hollow center of
the wall, the stack effect, or pressure from the gas entering below, pushes radon upward through the hollow centers. Research by the Pennsylvania Department of Environmental Resources found that the highest
concentrations of radon generally accumulate in the bottom several feet of a hollow block wall and seep out
through cracks and into the basement air.

If the house is well constructed, the top row of concrete blocks in those walls will be covered or somehow
plugged. Sometimes, however, a builder will overlook
this detail and the top row is uncovered. When there's
radon about, an uncapped wall simply funnels radon all
the way up its hollow center and right out the top, pipe-
lining huge amounts of radon gas into the basement.

Worse, the gas may push right up through spaces between the floorboards above the block wall and directly into the living space upstairs.

Any type of basement wall—masonry block, poured concrete, stone, or wood—that has water leaks also has potential radon leaks. The radon in the ground around the wall simply dissolves into the water there and is carried in along with the moisture. Thus, a wet basement is often a radioactive one as well.

In addition to that potential pathway, even the smallest crack or opening in a basement or foundation slab is as good as an engraved invitation to radon gas. Studies have shown that a crack of only 0.5 millimeters—about half as large as the head of a pin—is enough to create a convective current that will bring radon indoors. If there are enough cracks and holes in your basement, it may be no better at protecting you from radon than if it were bare soil.

THE OFFENDERS

You've really got to play detective to find radon's paths into your basement. The following are some possible clues, depending on how your house is built. In order of their importance, these are the most common entry routes into basements and slab-on-grade foundations: sump-pump openings, floor cracks and joints, basement floor/wall joints, basement floors of uncracked concrete, and basement floor drains.

SUMP-PUMP OPENINGS

If you have a basement that floods after heavy rains, a sump pump can be a godsend. These electric units have a

special float that senses a rising water level and turns on
the pump to suck water out of the basement and into a
storm drain, sewer, or onto the ground outside.

A sump pump is usually installed in a 24-inch-wide
pit with a gravel base that's dug below the basement

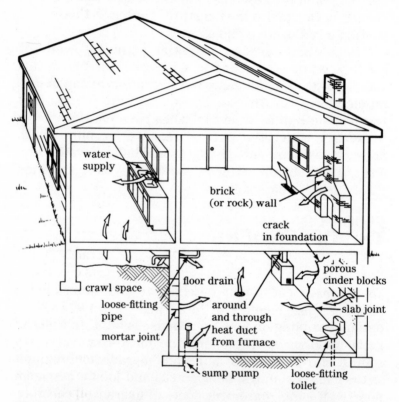

water
supply

brick
(or rock) wall

crack
in foundation

porous
cinder blocks

crawl space

floor drain

loose-fitting
pipe

around
and through
heat duct
from furnace

slab joint

mortar joint

sump pump

loose-fitting
toilet

*These are some of the ways radon can get inside a house.
Most of the entry routes are in the basement, since that's the
part of the house with the greatest surface area exposed to
the surrounding soil. The most common pathways are
through cracks and spaces around pipes, sump holes, floor
drains, and the joint between the floor and walls. The gas
can also enter the house dissolved in the water.*

floor. And therein lies the problem. The hole provides an ample conduit between the soil under the foundation and the inside of the basement. So it's not surprising that radon tests often show the highest concentrations of radioactivity in the area surrounding a sump pit.

Some sump pumps have covers, which can at least partially block the radon pathway. If yours has one, you should make sure that it's always on and that it fits tightly.

Similar problems are caused by downward flushing toilets installed in basements. Although the base of the toilet may look like it's creating a good seal with the basement floor, it probably isn't. And chances are that the holes broken through the concrete for the toilet's water and sewer pipes are larger than the pipes themselves and could be letting radon in from below.

FLOOR CRACKS AND JOINTS

The next most probable entry routes are cracks in the floor of the basement or tiny gaps at the joints of various parts of the foundation. For example, the "cold" joints formed by more than one concrete pour when the foundation was being built may not be completely sealed together, leaving gaps that can transport radon.

Other routes are the expansion joints sometimes left in the foundation to allow for movement of the concrete, spaces around loose-fitting sewer or water pipes entering the basement, basement sink drains, utility entrances, and hot water conduits coming through the slab.

Remember, what may be a barely visible crack or a concrete seam that looks like it was sealed with joint compound may still provide enough room for radon gas to seep in.

BASEMENT FLOOR/WALL JOINTS

The cold joint where the slab meets the basement walls is probably the most common entry route for radon but lets less radon into the house than floor cracks and other joints in the floor itself.

Occasionally there's even an intentional gap of a few inches between the floor perimeter and the wall, for drainage. These "French drains" can be a major entry route for radon.

BASEMENT FLOORS OF UNCRACKED CONCRETE

Although an uncracked foundation slab of concrete is a good radon barrier, it's not impermeable to the gas. Hardened concrete may look solid enough, but in fact it's somewhat porous. Concrete is made by mixing cement paste and crushed rock with water, and when the water evaporates it leaves small air spaces. Thus, a slab of cured concrete may actually contain more than 20 percent air. Radon can fill those air spaces and move through a slab by diffusion. It's a slow and inefficient process, and the amount of radon passed on by a given square inch of sound concrete is probably quite small. But there are a lot of square inches in the average concrete floor, and the overall diffusion rate may contribute to indoor radon problems.

BASEMENT FLOOR DRAINS

Basement floor drains usually connect directly to a sewer pipe or weeping tile underneath the basement. Most houses have at least one, to siphon off water from leaks or washing machines and sinks. The drains make a direct link from the interior of the basement to the ground below and may account for as much as 10 percent of the total air leakage into the average house and as much as 50 percent in a tightly constructed one. And along with cold outside air, they can let radon gas in. Even a water trap in the drain won't always deter radon; the gas is water soluble and can pass through the trap.

Then why are these drains last on the list of radon sources? Because, although they certainly let in more radon per square inch than unbroken concrete, they are relatively small (6 to 8 inches in diameter) in comparison to the total floor area.

OTHER ENTRY ROUTES

Of course, all of the above radon sources assume some fairly modern construction techniques, and not all houses meet that description. For example, some old houses, usually those in rural areas, may have bare dirt or gravel floors, or large spaces, such as a root cellar, that are unpaved. "If a house or part of a house is built over a dirt floor basement, it may be exposed to excessive amounts of radon," reports Conrad Weiffenbach, an assistant professor of physics at the University of Maine. In two given houses, each with an equally strong source of radon underneath, the one with a bare earth basement is likely to have 10 to 100 times more radon in the living area than one with a 4-inch-thick concrete slab over the soil.

A greenhouse or sunspace attached to a house could also be an inlet for radon, particularly if it contains large amounts of soil or, worse, has a dirt floor. However, ventilation rates in these glass structures are usually higher than in the rest of the house, so even if more radon is getting in, a lot gets swept out again in the ventilated air.

RADIOACTIVE SHOWERS

Eleanor LaCombe, of New Gloucester, Maine, works as a researcher at the Maine Medical Center in nearby Portland, studying the health effects of exposure to radon. She takes her work seriously enough that she decided she had better test her own house to see how much radon *she* is exposed to. The results of the tests shocked her: They showed that the radon levels in the drafty old farmhouse were 25 times higher than normal. The problem was traced to her water supply, which was contaminated with radon gas by the granite rock below the house.

"I don't want to panic, but I don't like the idea [of being exposed to so much radon] at all," she says. " Even if it's only when I'm taking a shower."

Like two-thirds of the homeowners in America, LaCombe gets her water from a well. And like about half of those wells, hers is near enough to a source of radon to allow the gas to become dissolved in the water (there are trace elements of radon in all water). Every time she turns on a tap, does a load of laundry, or uses any water in her house, LaCombe releases unhealthy amounts of radioactivity into the indoor air.

Radon in water is a serious concern. Christian Rice, a spokesman for the United States Environmental Protection Agency (EPA), says that as many as 1,000 Americans die each year as a result of breathing radon from water sources. In Maine, one study published by the New England Waterworks Association established a link between high radon concentrations in well water and an increased incidence of lung cancer.

Groundwater that contains radon can add substantially to the amount of radon in a house. According to estimates by the EPA, 2 to 5 percent of the radon exposure in the average home is caused by radon carried

LIBERATING RADON

If there's radon gas dissolved in your water supply, your daily household activities will release varying amounts of it into the indoor air. The efficiency with which the dissolved radon will be liberated from the water depends a lot on the water's temperature, how it's used, and how many gallons are used. Here are the average number of gallons used in, and the radon transfer efficiencies of, some normal activities:

Use	Daily Amount (gal)	Transfer Efficiency (%)
Bathing	40	47
Cleaning	2.6	90
Doing laundry	34	90
Drinking and cooking	8	30
Flushing toilet	96	30
Showering	40	63
Using dishwasher	14.5	90
Total	235.1	

Source: University of Texas School of Public Health.

inside via the water supply. But in some homes, like the one in Maine that has measured radon levels in its water more than 1,000 times higher than normal (the highest amount ever recorded) or the one in Pennsylvania that was 675 times higher, radon from water can account for nearly half of a home's contamination, the agency says.

Radon is water soluble and readily dissolves in the underground water that feeds private wells and some city water supplies. As the water splashes out of our faucets, showers, or washing machines, part of the dissolved radon is liberated into the air.

An EPA study found that doing the laundry with radon-contaminated water releases the most gas into the air, accounting for 25 percent of radon from water sources in the average four-person household. Flushing the toilet was a close second at 24 percent, followed by showering at 21 percent.

The highest amounts of radon seem to be found in drilled wells up to about 150 feet deep, report researchers from the University of Maine who have been looking into the problem. Beyond that depth, they say, radon levels start to decrease. That finding could be due to the fact that shallower rock has more cracks and is more porous than deep rock, and because water in deeper wells stays around for as much as five days before it is

RADIOACTIVE MATERIALS: AVERAGE CONCENTRATIONS OF URANIUM IN BUILDING MATERIALS

Material	Uranium (ppm)
Cement	3.4
Dry wallboard	1.0
Granite	4.7
Limestone concrete	2.3
Sandstone	0.45
Sandstone concrete	0.8

Source: David I. Poch, Radiation Alert. *(New York: Doubleday, 1985).*

pumped to the surface, giving some of the radon time to decay. Numerous studies have shown that wells in areas of granite rock are the most likely suspects for high radon readings.

Surface water almost always has far less radon than water from wells, probably because the greater surface area exposed to the air causes much more of the radon to escape into the atmosphere.

If you're served by a public water system, there seems little reason to worry about radon in your water. The aeration process that most treated water is subjected to removes much of the radon at the plant, and more will be lost in the pipes on the way to your house. A study at the Lawrence Berkeley Laboratory (LBL) in California found that well water contributes 20 to 40 times more radon to houses in the United States than does water from public supplies.

BUILDING WITH RADON

Some houses are sabotaged not by the radon in the ground around and under them, but by the very materials used to build them. Although in most homes the radon given off by building materials isn't a major source, it can contribute to an existing problem. And in a few instances it can cause significant radon pollution.

Some building materials, such as wood, most metal, plastic, and insulation give off almost no radiation. Others, primarily stone and masonry products made in areas with naturally high radium content, produce varying amounts of radon. Studies from all over the world have measured the radon content of building materials. In order of severity, the building materials most often associated with indoor radon are granite, concrete, clay brick, and gypsum.

GRANITE

Granite block is a greater potential source of radon than any other building material, due to the fact that it usually contains relatively large amounts of radium. Luckily, the stone isn't used much in home construction, except occasionally as decoration on the exterior of a fireplace or to cover a foyer floor. Some older houses in New England and elsewhere are built on foundations of cut granite block, but the amount of the stone used relative to the size of the house makes it doubtful that the granite would contribute a lot of radon to the air inside.

Some office buildings, apartments, and college dormitories have been built from granite, and that might be another story. Very few large multi-unit or commercial buildings have been tested for radon, so no one knows for sure, but it's entirely possible that the rock they were built from could be polluting the air inside with radon. In Sweden, people who live in stone houses have been shown to have higher lung cancer rates than people who live in wooden homes.

CONCRETE

The radioactivity of concrete varies widely, depending on the radium content of the materials used to make it. Swedish research indicates that the radium content of the cement fraction of concrete is quite small, but the aggregate part (sand and gravel that's usually supplied locally) may boost the radon output of the concrete if the regional soils are naturally high in radium. When scientists at LBL studied 100 concrete samples from ten parts of the United States, they found that the radon gas given off by the materials varied by as much as a factor of 10 from sample to sample. Overall, though, the LBL researchers estimate that concrete accounts for no more than 10 percent of the average home's indoor radon.

A big exception is the concrete into which phosphate slag was incorporated. This was done in some parts of the country, particularly southeastern Idaho, in the early 1960s, apparently before anyone was giving much thought to radon. An estimated 74,000 houses were built with the concrete, which has since been proven to have a very high radium content.

When fly ash from coal-burning power plants is added to concrete mixtures, even in small percentages, it can significantly raise the radium content of the cement.

CLAY BRICK

Again, the radium content of bricks depends on that of the soil they were made from. So far, no one has undertaken an extensive study to measure the amount of radium in bricks throughout the United States, but tests conducted on bricks in Europe and the USSR turned up some fairly high readings. However, the tightly packed nature of clay makes it more difficult for radon to diffuse through, so while the radium content of a brick may be elevated, the amount of radon gas that escapes will probably be quite small.

GYPSUM

Ordinarily, building materials containing gypsum, such as wallboard, are low in radium, containing only about one-fourth as much as granite, on average. But there are occasional exceptions. For instance, phospho-gypsum, a by-product of phosphate mining, carries amounts of radium hundreds of times higher than those in granite. It has found its way into wallboard and was once used in making insulation for some houses in the state of Washington.

NATURAL GAS

As we've seen, natural gas can also be a carrier of radon, although most of the radioactivity decays before it reaches users. It's possible that some homeowners who live near gas wells could be receiving larger-than-average amounts of radon in their gas supply. But even so, it's doubtful that much radon is actually getting into their indoor air. In the old days, unvented gas heaters and fireplaces were common (no wonder the life expectancy was shorter back then). Today, nearly all gas appliances, with the exception of some cooking ranges, are vented outdoors.

SUSPECT HOUSES

If you look anywhere but in one of the few absolutely homogenized developments of the 1950s and 1960s, it is hard to find two houses that are exactly the same. Sometimes the differences among houses is slight—a small variation in room layout or fewer windows—but these differences can change not only the look, but the dynamics of each house: such as the way air moves into, out of, and through it.

How a house is designed, built, modified, and heated can all have a major effect on how much radon the people living in it are exposed to. That's one of the reasons it's so hard to predict which houses will have problems without first testing them. But so far, researchers working in the field have been able to come up with some rules of thumb about how radon travels through a house once it's gotten inside. Here are some of the findings published by the United States Department of Energy (DOE) and other government agencies:

Radon levels are almost always higher in lower parts of a house. Since the soil and rock under a house are the main sources of radon, and basement or slabs the main entry points, it stands to reason that radon levels will be higher in the sections of the house closest to the source. Indoor radon concentrations usually show a decrease from the basement to the first floor and again from the first floor to higher floors. According to Dr. Cohen's extensive test results, basements have concentrations two to three times higher than that of first-floor living rooms. Other tests have shown that the levels on the second floor are likely to be reduced by half again.

It would follow that people living in the upper floors of high-rise apartment buildings have little to fear from radon. And that's *usually* true. But in buildings where elevator shafts or air ducts connect higher floors with the basement or subbasement, the rules change. A study of two high-rise buildings in Birmingham, England, one 11 stories, the other 17 stories high, found that radon levels weren't necessarily lower on the upper floors. Apparently, an open shaft that begins in the basement can act as a conduit to channel radon gas all the way up to the penthouse.

How you heat your house can make a difference. The National Council on Radiation Protection says that in homes with electric baseboard heating systems, radon levels increase in the winter because the house is sealed up and radon gas has fewer avenues of escape. The opposite is true in warm climates, where the house is more likely to be sealed up in the summer to keep air conditioning bills down.

Houses with gas and oil heating systems sometimes show a decrease in radon concentrations in the winter as basement air (and radon) is sucked into the furnace during combustion and then is carried up and out through the chimney. However, in most houses heated with a furnace, the air for combustion comes from inside the basement. As the furnace or boiler draws that air in, a slight negative pressure is created in the room, a vacuum that can draw more radon in from outside.

Floor layout affects where radon travels. Radon is a gas, so you can expect it to move around through a house much the same way air does. A barrier that will divert the flow of an air current will also redirect radon. Thus, interior design can determine how radon gas is distributed within a house.

For example, walls and interior partitions restrict the flow of radon. Doors, stairwells, kitchen/dining room pass-throughs, and other interior openings act as passageways for radon to move throughout the house.

Even ceiling height can make a difference. According to a report from Sandia National Laboratories, high ceilings may cause air to stratify in layers, creating a build-up of pollutants in the stagnant sections.

ARE ENERGY-EFFICIENT HOMES DANGEROUS?

Before the mid-1970s most of us spent very little time worrying about the energy efficiency of our houses. When heating oil was 15 cents a gallon, there wasn't much reason to be concerned if a few gallons were being wasted every day.

Then along came the Arab Oil Embargo. Prices shot up and supplies plummeted. And most sensible people began looking for ways to make every expensive gallon of fuel go farther.

One word that millions of us learned the importance of was *infiltration*—the amount of cold air leaking into our houses through cracks around windows and doors, penetrations through walls for electrical wiring and plumbing, and the countless tiny openings in the typical building shell of the time.

The American Council for an Energy-Efficient Economy has calculated that if all the leakage sites in the average house built prior to the 1970s were added up, they would make a hole the size of a basketball inviting in the frigid winter air.

So we sealed up our houses, adding weather stripping, caulking, and insulation. And we built efficient new houses that were twice as airtight as they used to be. The result has been energy savings worth billions of dollars a year to Americans.

But there are those who are concerned that these savings bring with them a hidden cost. Scientists and health officials around the country have expressed fears that as we tighten up our homes and reduce the infiltration of fresh air from outdoors, we're eliminating escape routes for radon and could be making bad indoor pollution problems worse.

Some early research indicated that that was indeed happening, touching off a debate that has yet to be completely settled. In one study conducted in New York State and reported in the journal *Health Physics*, radon levels in 14 energy-efficient houses were said to be twice as high as those in 16 leakier houses. In another study, an experimental house showed radon levels 25 to 30 percent higher after it was tightened up to reduce infiltration. Dr. Cohen's results show that well-weatherized houses *seem* to have" 40 percent higher radon levels than poorly weatherized houses.

Other researchers point out that more than half a dozen studies in the United States and Canada have since failed to show a significant link between building tightness and indoor pollution levels. They argue that in some early studies, the determination of whether a house was energy efficient or not was reached by asking the owners on a questionnaire. That, they say, is not a very accurate way to prove how tightly built a house is.

Several types of energy-efficient construction continue to be under suspicion, though. Earth-sheltered and underground homes have a larger surface area in contact with the soil, which means more places for radon to enter. "We've yet to see an underground house that doesn't have some radon in it," says B. V. Alvarez, president of Airchek, a radon testing company. Based on the results of several tests, "it appears that for areas in which conventional homes tend to exhibit higher than average radon, the use of earth sheltering may cause significant increases," the DOE now says.

Warning flags are also being raised about "superinsulated" houses, so called because they are built particularly tightly and contain far more insulation in walls and ceilings than do ordinary homes. In New York State, for example, two researchers discovered unsafe radon levels in a superinsulated house that they had anticipated would test low. Further research revealed that the holes inside the concrete building blocks used in construction had been filled with sand to enhance their ability to store solar heat. Even though sand usually doesn't contain a lot of uranium, in this very tight house it gave off enough radon to raise the indoor concentration to more than twice the safe amount.

The bottom line, according to Anthony Nero, of LBL, is that some energy-efficient houses have been found to be badly contaminated with radon, but some leaky ones have, too. Tightening up a house may aggravate an existing radon problem, but it usually won't create trouble where none existed before. "If you have radon and you tighten up a house, you might increase the indoor radon levels modestly—say by 10 or 20 percent," he says. "In almost no case can I conceive of it making a big change, except maybe if you built an extremely tight house, and there are only a few of those."

So, simply turning up the thermostat and letting cold air seep into your house, in addition to being an expensive option, is no guarantee that you won't be exposed to radon. There are many other factors at work, including, and probably most importantly, the strength of the radon source beneath the house and the permeability of the soil. And as you'll see later in this book, you can have any kind of house you want, anywhere you want, and still live radon-free.

4

RADON AND YOUR HEALTH

The good news about radon is that most people are not exposed to enough of it to seriously harm their health. The bad news is that for the rest, primarily those who have lived or worked for years in an environment heavily contaminated with radon, the risks appear to be great. For many years medical researchers have puzzled over the the causes of the large number of yearly cases of lung cancer not linked to smoking. Now, it appears that radon may be the culprit. The National Cancer Institute has singled out radon exposure as the number one cause of lung cancer among nonsmokers, accounting for as many as 30,000 deaths every year in the United States.

Others say the number may be even higher. "I think radon could be responsible for 40,000 cases of lung cancer a year in people who don't smoke," says Richard Toohey, of the Argonne National Laboratory. Either number would place radon exposure in the top ten causes of death in this country.

Perhaps what's most disconcerting about the health effects of radon poisoning is that there are no acute or short-term symptoms. If there were, the problem would be less frightening, because the sickness would warn of unsafe levels of the gas. Unfortunately, the damage done by radon is never immediately apparent. The Watras family and others have lived in houses that were spectacularly radioactive for months, even years, totally unaware that anything was amiss.

Radioactivity from radon does its damage quietly, injuring the body a few cells at a time. Only after 20 or 30 years, and even longer, may the disease it causes make itself known.

"Radon is a cancer time bomb," says Robert Yuhnke, an attorney for the Environmental Defense Fund. "As we build more and more houses on land contaminated with radon, we're exposing more people to it. There's usually a minimum 20-year latency period before cancer strikes. Ten years down the road, we may be seeing an even higher incidence of [radon-related] cancer than we do now."

We're all exposed to some natural radioactivity every day—from the sun, cosmic rays, radon in the outdoor air, and even some of the food we eat. The levels of this "background" radiation are tiny and so is the harm they cause us. For example, breathing the amount of radon in the air outside every day is about as dangerous as smoking seven packs of cigarettes *a year*. The average skiing trip will put you at greater risk of death.

The air in the average house contains up to ten times more radon than the air outdoors. You're up to maybe 30 packs a year, not great, but still nothing to lose sleep over.

But for an unlucky number of people—no one knows yet exactly how many—life at home means daily exposure to extremely high levels of radioactivity. Houses that test hundreds of times higher than average are not uncommon. And in Pennsylvania, the Watras family would have been safer smoking 2,000 cigarettes a day than they were living in their house before it was "fixed."

Daily contact with large amounts of radioactivity is unhealthy for humans. Although radon is a "natural" radioactive substance, the body recognizes little difference between it and radioactivity from man-made sources such as nuclear power plants or atomic fallout. Physiologically, the effects are very similar.

THE ENEMY WITHIN

As we've seen, the decay of radon gas produces a series of short-lived daughters (bismuth, lead, and polonium), which are all radioactive. About 90 percent of the radioactivity emitted by radon and its daughters is in the form of alpha particles, the most dangerous type. Numerous studies have proven that alpha radiation is 10 to 30 times as serious biologically as an equal dose of beta particles or gamma rays. Because of the high percentage of alpha radiation emitted from uranium and radium, the International Commission on Radiological Protection has classified them as "by far the most toxic of all radioactive trace elements."

Despite their ability to raise havoc with living cells, alpha particles are low-energy radiation with little penetrating power. Unlike X rays, a type of gamma radiation, alpha particles can't pass through your skin and enter your body that way.

But radon's daughters do find their ways to invade the body. They sneak in hidden in food and water, and in the air we breathe.

RADON A LA CARTE

Scientists say the food some of us eat may contain radon. A carrot grown in naturally radioactive soil will pick up some of the radon as it grows. For instance, researchers who tested 50 plant species for the Bhagha Atomic Research Center in Bombay, India, concluded that " . . . the levels of radioactivity observed in plant species from high radiation areas were considerably higher than the corresponding plants from areas with normal background radiation."

In general, the higher the radioactivity of the soil, the more the food is likely to carry. The National Council on Radiation Protection and Measurements (NCRPM) says the radon concentrations in vegetables directly correlate with the amount of radioactivity contained in the soil they were grown in. Root crops and cereal grains tend to absorb more radon than fruits and leafy vegetables. It's even possible for cattle that graze on contaminated plants to concentrate radioactivity in their organ meat.

Most of the food we eat is grown elsewhere and shipped long distances, so it's safe to assume that much of the radioactivity it may have contained when picked will have decayed to much lower levels by the time it reaches the dinner table. However, some locally grown produce in areas of high natural radioactivity may still contain significant amounts of radon when it's brought home.

Radioactive food may also be a problem in some of the western states, where uranium is mined and milled. According to Ralph Nader and John Abbotts, in *The Menace of Atomic Energy*, tests of the Animas River downstream from a uranium mill in Durango, Colorado, have shown high levels of radium. They say the drinking water may be contaminated with unhealthy amounts of radium and that radioactivity in the river can be concentrated in the food chain: Plants in the river had higher amounts of radium than the water surrounding them, and fish that eat the plants had higher levels still. The authors cite one study that found that the flora and fauna in the river contained 100 to 10,000 times the concentration of radium in the river water.

WATER FIT TO DRINK

Drinking radon-contaminated water, or eating food that contains the water, can also increase a person's radiation dose. Usually, the amounts of radiation picked up from drinking water are small, since average radon levels in water are relatively low. But in some houses, where well water radon measurements show counts thousands and *hundreds* of thousands of times higher than normal, occupants could be drinking a harmful dose of radiation every day.

The radon daughters in water won't have enough energy to pass through the stomach walls, but radon gas can be burped out into other parts of the body. According to a United Nations report on the effects of radiation, some of the radon ingested in water is eliminated by the body, but residual amounts can remain in the stomach, liver, kidneys, fat, and bone marrow.

IN THE AIR

By far the largest internal radon exposure, and the most dangerous, comes by breathing radon. Inhaling radon gas itself is relatively harmless, since the gas is breathed back out again before any significant decay takes place. (Radon has a radioactive half-life of 3.8 days; the average breath stays in the lungs only about 17 seconds.) What can be a threat to health are the radon daughters that are formed directly in the air from radon gas. The air in a contaminated house may be dense with thousands of radon daughters, continually decaying.

Most of the electrically charged radioactive daughters—the heavy metals lead, bismuth, and polonium—have an electrostatic attraction that causes them to act as magnets to other particles in the air. They'll attach themselves to just about anything in the house: walls, furniture, clothes, dust, or particles from wood and cigarette smoke. This attachment is known as *plating out*. The remaining daughters, usually less than 10 percent, remain unattached to anything.

In walks a person and breathes the radon-laden air. The daughters, attached and unattached, are inhaled into the lungs and about 30 percent of them come into contact with the air passageways and stick to the mucous lining of the respiratory tract. Large particles are stopped in the upper part of the bronchial tract while smaller ones travel deeper into the lungs. (The lungs have built-in mechanisms to clear themselves of dirt and dust, but the process may take a day, by which time radon daughters have already spent their radioactivity.) Some particles may also lodge in the nose and pharynx, and from there they may be absorbed into body fluids, breathed back out, or swallowed into the stomach. Radioactive particles can even find their way into the lymphatic system.

The radon daughters that are unattached, when inhaled, are nearly 100 percent efficient at adhering to the lungs and can be more harmful than those already clinging to dust particles. The National Council on Radiation Protection estimates that this unattached fraction can deliver up to 40 times more radioactivity to the lungs than the portion attached to particles in the air, mainly because of their better adhering ability. (Medical researchers say the lung dose from unattached particles is highest among people who breathe primarily through their mouths; nose breathers get some protection from the filtering action of the nasal passages.)

HOW RADON DAMAGES CELLS

Once deposited in the lungs, or the stomach and other internal organs, radon daughters continue to decay, transmitting radiation to nearby cells. Whether the radon is drunk, eaten, or inhaled, the damage that alpha radiation can inflict on living cells is virtually the same.

Unlike beta and gamma radiation, which can penetrate far into the body and disperse their radiation energy over a wide area, an alpha particle, which is actually a helium atom, travels a relatively short distance through human tissue and concentrates its radiation on a small area of cells.

Moving at more than 4 million feet per second, the relatively large alpha particles can smash past molecules in their path, knocking off a great many electrons, or ions, as they pass. To help people understand this process, Charles Hess, a physicist at the University of Maine, uses the analogy of a bowling ball (the alpha particle) hitting marbles (electrons). "Alpha particles lose speed as they pass through matter and knock electrons off molecules of such matter. This is similar to a situation where a bowling ball would be slowed down by hitting marbles suspended on strings. Because the marbles have so much less mass than the bowling ball, the ball will only be stopped after hitting a great many marbles."

In air, alpha particles will slow to a halt after traveling 2 or 3 inches. In liquid or solid matter, such as body tissue, the higher density will stop the particle after about 1 thousandth of an inch. Along the way, the disruption these "bowling balls" cause at the nuclear level is the heart of radiation damage.

A large enough blast of alpha energy can disrupt enough electrons to kill cells. If the exposure were large enough, so many cells would be destroyed that the radiation dose would be fatal. But more often, a moderate

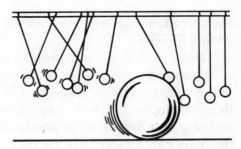

This picture of a bowling ball rolling through a field of marbles suspended on strings is meant to show how a large alpha radiation particle (the bowling ball) can disrupt many electrons (the marbles) before the repeated collisions slow the ball to a stop.

exposure will kill only a few cells. That's not good, but thousands of cells die every day and the remaining ones are able to reproduce themselves to replace the losses without serious harm to the body.

In the long run, it may be healthier to kill a few cells with radiation than to only damage them. It appears that low-level doses of radiation, not strong enough to actually destroy cells, are much more dangerous. Small, repeated exposures to alpha particles can alter cells instead of killing them. And the alteration of a single cell over time may lead to cancer.

No one knows for sure how or why cancer cells erupt into uncontrolled growth, or why there is often a 20-year or more lag time between exposure to radiation and the onset of the disease. But there are theories.

One has to do with the way that radiation, and particularly alpha radiation, can harm DNA (deoxyribonucleic acid) molecules, the substance in chromosomes that stores genetic information and the pattern from which all cellular components are synthesized.

Heavy alpha particles plowing through DNA in a cell nucleus will often break both strands of the double-stranded molecule. Whereas less dense beta particles or gamma rays may break a DNA molecule in one spot, or not at all, as they pass through, a very dense, ionizing alpha particle is likely to break the DNA strands in three or four places.

DNA molecules are broken all the time, by toxic chemicals, air pollutants, and even normal body heat. After such "routine" damage, DNA molecules can produce enzymes that have the ability to remove the affected pieces and heal the breaks without damage to the precious cargo of genetic codes. If the break is repaired correctly, the DNA molecule is virtually as good as new.

But the more extensive injury caused by an assault from alpha radiation may overwhelm a DNA molecule's capacity for self-repair. When the breaks are multiple, it's more difficult, and often impossible, for the DNA molecule to heal itself properly. A broken section may fail to rejoin with the main part and be lost. Or, when two breaks are close together, the broken ends from one of the breaks may latch onto the ends of another break and recombine the wrong way. When a repair isn't complete, or is done incorrectly, changes in the chemical makeup of a gene may be altered and the genetic material it carries destroyed or scrambled.

What does that mean? Probably not much in the short term. But over time it may lead to serious changes in the cell that the DNA governs. The problems come when a DNA molecule reproduces itself in order to create a new cell. Cell division is the way our bodies regenerate themselves every day, creating new cells to replace dead or dying ones. If a DNA molecule is not repaired, or improperly repaired, it will still produce an exact duplicate of itself, flaws included, over and over again. In time, many cells are created that carry incorrect genetic messages.

It's not entirely clear how that process may lead to diseases like cancer. One theory holds that imperfectly repaired radiation damage may interfere with a cell's internal growth regulation. Ordinarily, cells divide only when needed, and produce only as many new cells as are necessary. But it's possible that when DNA molecules are injured by ionizing radiation, the regulatory mechanism that keeps cell division in check goes haywire and cells proliferate uncontrollably, leading to a tumor.

Another scenario is that radiation damages DNA by creating toxic chemicals within cells. As alpha particles pass through cells, knocking electrons around, other molecules may catch the free electrons and change into new chemicals. When the oxygen contained in cell fluid captures a floating electron, it turns into a highly toxic "superoxide radical" that can damage the cell and the DNA it contains.

Finally, there's a possibility that prolonged exposure to low levels of radioactivity may cause defects in the body's immune system. When that happens, the body may not be able to muster its defenses to detect and fight off growing cancer cells.

One frustrating aspect in the investigation of how radiation causes cancer is the fact that there is almost always a lag time of at least 10 years, and sometimes more than 40 years, between the initial exposure to radioactivity and the onset of the disease. (About half of all cancer patients are 65 or older when first diagnosed.) Just what goes on during this "latency period" is a mystery. Considering that there are about 1 billion cells in $\frac{1}{32}$ ounce of human tissue, it could simply take that long for a cell carrying a distorted DNA molecule to reproduce the millions of times it would need to in order to grow large enough to be a noticeable cancer.

RADON AND LUNG CANCER

Medical researchers don't yet have all the answers about the origins of cancer, but they are sure of one thing: Radon can cause it. As a report by the National Research Council put it, "There is no doubt that radon in sufficient doses can produce lung cancer in man."

The reason everyone seems so sure on that point is the body of evidence acquired over many years, at a cost of thousands of lives. The "subjects" who proved beyond a shadow of a doubt that radon causes lung cancer were uranium miners.

Actually, hints that something was very wrong down in the mines were surfacing *centuries* before anyone even dreamed about atomic power. In 1576, a book about the mining of a mineral called *pitchblende* in the Erz Mountains of Germany mentioned the "mountain sickness," a chest disease that was killing many of the miners.

No one seemed to take much notice at the time, because it took more than 300 years for anyone to take a serious look at the problem. In 1879, a medical study finally identified the source of mountain disease as lung cancer. The researchers also presented an astounding statistic: 50 to 75 percent of the men working in the dusty, poorly ventilated mines of southern Germany and Czechoslovakia were dying of the disease.

The discovery of radioactivity was still decades away, and it wasn't until the late 1930s that someone suggested that the cause of all those deaths in the Erz Mountains was radon gas emitted from the pitchblende, now known as uranium, that was mined there.

Incredibly, that discovery didn't lead to great improvements in mine safety. It wasn't until the 1940s, after an Atomic Energy Commission study found that radon levels in mines in the United States were *higher* than in European mines, that the first steps were taken to remedy the situation. By that time, the lung cancer rate among miners was still over eight times higher than that of the general population.

Better ventilation systems were installed in many mines to help carry off the clouds of dust, radon, and diesel fumes that had been an accepted part of the miners' daily work life. And as mine conditions continued to improve in the 1950s and 1960s, the lung cancer epidemic that had plagued miners for hundreds of years started to decline.

Decline, but not disappear. In 1976, a surprising study of the Czechoslovakian mines appeared. It showed that as radon levels in the mines had begun to fall in the 1940s, after conditions improved, the lung cancer rate among uranium miners remained higher than for the general public. In fact, the excess number of lung cancers had fallen by nearly *the exact same percentage* as radon levels in the mines had dropped. In other words, lowering the amount of radon in the mines didn't eliminate the threat of lung cancer, it just lowered the risk accordingly.

A final study comparing radon's effects with those of equal amounts of radiation from X rays and A-bombs showed that the lung cancer risks were the same for radon as for the other two. That, according to University of Pittsburgh physics professor Bernard Cohen, was the final piece of evidence the scientific community needed. "This cemented the case," he wrote in *Consumer's Research* magazine. "With all these data, radon-induced lung cancer has become one of the most quantified health effects of radiation."

Although the price was needlessly and tragically high, the medical community now has a great deal of information about how exposure to various levels of radon gas impacts on lung cancer rates—and how those levels might affect people living in radon-contaminated homes.

When a person breathes in radon-contaminated air, as many as one-third of the dust particles with daughters attached, and most of the unattached daughters, may lodge in the lungs. Some travel deeply into the lungs, but most get stuck in the moist upper bronchi, the two main branches that lead from the trachea (windpipe) to the lungs.

Just at the surface of these large bronchi is a layer of tissue 1.5 thousandths of an inch thick (about a third the thickness of the paper this page is printed on). Of all the parts of the human body, this layer, called the *epithelium*, is one of the most sensitive to radiation. That's because the epithelium is made up of *basal* cells, a type that reproduce much more rapidly than other types of cells.

Trapped in this delicate tissue, the short-lived radon daughters, especially polonium-218 and polonium-214, with half-lives of only 3 minutes and 0.0002 seconds respectively, deliver their load of alpha particles to the tissue around them. The range of these particles, you'll remember, *is 1 thousandth of an inch or less,* so nearly the entire burst of radioactivity is concentrated in the outermost lung tissue. "Once inside the lungs, the alpha particles from radon have an exquisitely placed target for their dose of radiation," notes Dr. Naomi Harley of New York University School of Environmental Medicine.

If the theory holds true that exposure to low levels of radiation can cause DNA mutations, and if it then follows that the duplication of those mutations can lead to tumors, then it's not surprising that repeated doses of radon daughters to the fast-reproducing cells of the lungs cause cancer. The more damaged cells there are in the lungs, the greater the chance that one begins the uncontrolled growth process that becomes cancer.

THE LEGACY OF THE MINERS

At first look, it's tempting to dismiss the studies linking lung cancer to radon in uranium mines as irrelevant to homeowners exposed to radon. After all, a house is not a uranium mine. In the worst of the mines, radon levels were thousands of times higher than in the average house. Workers were breathing so much radioactive dust that they could literally set off a Geiger counter simply by blowing on it! This leads us to the following question: Isn't the medical community comparing apples to oranges when it uses lung cancer rates among miners to estimate the danger to people living in homes contaminated with radon?

Not really. Certainly the figures from mines don't make it easy to calculate exactly how many homeowners are at risk. For one thing, there were almost no female uranium miners, so it is impossible to guess how the same radon levels in mines would have affected women. (All things being equal, women have historically had a lower overall lung cancer rate than men.) And neither the very old nor the very young worked in the mines. So we are not dealing in an area that could be called a precise science, which is why estimates of the number of yearly cases of lung cancer from indoor radon range from 5,000 to 30,000, depending upon whom you ask.

Nevertheless, the statistics from miners are significant. Early miners exposed to extremely high doses of radon had an equally high incidence of lung cancer, yet safety steps that reduced the radon levels in the mines didn't eliminate additional lung cancers among miners. They only lowered the number correspondingly. In other words, a 10 percent reduction in exposure brought a 10 percent reduction in lung cancers.

Studies in Sweden have shown excess lung cancers occurring among miners exposed to relatively small amounts of radon, levels not unusual in homes in some parts of the United States. And Dr. Harley, who has done extensive research on uranium miners, says, "We've seen excess lung cancers in Canadian miners that were exposed to radon levels not that much higher than you'd find in many houses. And we've also seen houses that are many times more radioactive than any of the mines ever were."

"I really believe you can make an extrapolation between uranium miners and homeowners," she concludes. "A given amount of radiation in a mine will affect the

WARNING: RADON AND CIGARETTES CAN BE EXTRA HAZARDOUS TO YOUR HEALTH

Now cigarette smokers have still another reason to quit: Doctors say that smoking can make radon more dangerous, not just for the person holding the cigarette, but for everyone in the house.

Originally, the news was a little better. It actually looked for a while like cigarettes could *protect* the smoker from radon. The idea was that the thick mucous layer that forms in the throat and lungs of smokers absorbed most of the radiation dose from inhaled radon daughters.

Sorry! An 18-month EPA study from more than 70 houses found that smoking can make a bad radon problem worse. The researchers say that the more dust and particles in a home's air, the more radon daughters there will be floating around in it. Without a handy particle to latch onto, most of the radioactive daughters will end up decaying harmlessly on the walls and furniture.

Smoking, of course, produces *lots* of particles. It can double the number of daughters in the air, putting everyone who lives there at greater risk. So if you must

miner no differently than the same dose will affect a homeowner—only the homeowner will be exposed to it for a longer time."

She's referring to the fact that, while a miner spends maybe eight or nine working hours underground, many people, especially children, the elderly, and parents who don't work outside the home, are in the house for much smoke, you'll be doing your family a big favor if you do it outdoors, especially if your house has already been found to have a radon problem.

On top of all the other evidence, consider the fact that cigarette smoke itself is radioactive. It contains the radon daughters polonium, bismuth, and lead, perhaps from the phosphate fertilizers used to grow tobacco. "When you burn tobacco, you produce some extremely radioactive smoke that is very carcinogenic," says Edward Martell, of the National Center for Atmospheric Research. "Add that to the radioactivity that's already in a house from radon, and you've got problems."

There's also some indication that cigarette smoke may cause tumors to appear faster. Data show that the lung cancer rate among uranium miners who smoked more than a pack a day was *seven to nine times higher* than that of nonsmoking miners, and that the cancer appeared decades earlier than among miners who didn't smoke.

If you're a smoker, you should probably be worrying much more about that than about the possible risks of radon. Smoking is said to kill 110,500 people yearly, probably three times more than radon. On the other hand, if you quit, you can worry less about smoking *and* radon.

longer than that each day. (The average American spends 85 percent of his or her time indoors.)

It may be simply that there is no such thing as an entirely safe exposure to radioactivity. Chronic exposure to low levels of radon's alpha radiation can scramble DNA data in cells and cause cancer, whereas larger doses tend to kill cells outright. And when a cell is dead, the possibility that cancer can develop there no longer exists. As a radon report from the Harvard School of Public Health points out, "...the risk of lung cancer from lower doses may be as great or greater per unit dose than the risk at higher doses."

A national study to determine how radon exposure is affecting the health of homeowners would seem in order. But it won't be an easy undertaking. Cancer studies of large populations tend to be very complicated. For one thing, care has to be taken not to confuse cancers from radon exposure with those caused by smoking. Then there's our propensity for moving every few years. A person who dies from lung cancer in Florida, for example, may have been exposed to radon while living in a contaminated house in New Jersey before he retired to Florida. Medical researchers have to play detective, using death certificates only as a starting point to trace the life history of the victim.

Finally, the cancer rate from exposures to low-level radiation is tiny when you look at it in light of the total population. To prove any cause-and-effect relationship, you have to study *lots* of people. In order to demonstrate even a 1 percent increased cancer risk from radiation, researchers would have to investigate millions of cases.

The government is in the process of undertaking some limited studies of household radon exposures, but preliminary results won't be available until the early 1990s. In the mean time, the legacy of thousands of uranium miners is the best anyone has to go on.

OTHER HEALTH EFFECTS FROM RADON GAS

Lung cancer is the only *documented* disease linked to radon gas. It's the problem that has gotten, and will continue to get, most of the attention, because of the large number of people who could be at risk. But what is beyond those numbers? What other health problems might daily exposure to radon gas be contributing to?

"All we have now are the lung cancer data," says Dr. Harley. "Anything else that's happening is trivial compared to that. But if there is something going on, it will have to get more serious study sometime in the future."

Given the modest funding currently allocated for studying radon, that study may be a long way off. So we're left with little better than speculation, based mainly on different types of exposure to low-level radioactivity that a homeowner is likely to encounter, such as nuclear fallout. In any case, the possibilities are worth a look, even if the jury remains out on their validity.

STOMACH CANCER

Can drinking water laced with unseen doses of dissolved radon gas lead to cancer of the stomach and gastrointestinal tract? The numbers on this form of cancer in the general population are small—only a few thousand cases yearly, so the danger wouldn't appear to be too serious.

The stomach isn't considered to be highly sensitive to radiation, and radon ingested through water or food probably doesn't stay in the body for a very long time in any case. When attached to dust particles, radon daughters can lodge in the lungs, but dissolved in the stomach, they can't stick and will eventually be flushed out by the body's systems. However, considering that some of radon's daughters have half-lives of less than half an hour, the stomach and intestines could be receiving a residual radiation dose before the radon is eliminated.

The potential for stomach cancer from radon can't be completely ignored. There are clusters of geographic areas where death rates from gastric cancers are significantly higher than the national average, among them New England, the north-central Midwest, and north-central and northeastern New Mexico. In New Mexico, at least, research points the finger at the presence of uranium as a possible reason.

When Dr. Gregg Wilkinson, of the Los Alamos National Laboratory, tabulated stomach cancer cases county by county in the state, he found that the parts of New Mexico with significant uranium deposits "are also characterized by high mortality rates for gastric cancer" and that the counties without uranium had a lower incidence of the cancer.

Dr. Wilkinson's study is inconclusive, however, on whether the stomach cancers are the result of radon in drinking water and food, or the presence of trace elements such as arsenic, cadmium, and selenium, which are often found in the vicinity of uranium deposits.

Preliminary data from health records of uranium miners, whose water supply while working often came from a source that flowed through the radioactive ore, suggests that they also show an increased incidence of stomach cancer, although the effect was small compared to lung cancer.

The jury is still out, but it seems at least possible that drinking radon in water may be a factor in some cases of stomach cancer, especially among people who are ingesting the large amounts found in well water in some parts of the country.

In Pennsylvania, where some problem wells have been discovered, the official word from the Pennsylvania Department of Health is this: "There is a potential risk of stomach cancer from ingestion of radon-contaminated water." And in Maine, where some of the well readings have come up astoundingly high, a research report from the University of Maine says, "The health risk of drinking water with such levels of radon would undoubtedly be considered significant enough to warrant corrective measures."

BONE CANCER

The evidence that links radioactivity to other forms of cancer writes another sad chapter in the history of working people. Only this time it was mostly women who suffered.

In the early 1920s, a chemist found that when radium, part of the decay chain of uranium, was mixed with substances known as luminizers, the radioactive alpha particles caused the concoction to glow in the dark.

Enterprising inventors soon found a number of uses for the new discovery, including painting it on watch and clock dials so they could be read easily at night. Wearing a watch with a radium dial became an instant status symbol and an industry quickly grew up to meet the demand.

Dial painting was one of the few job opportunities for young women at the time and thousands went to work in the factories, the largest of which was the United States Radium Corporation in New Jersey. The workers were paid by the number of watches they completed every day, but their painting had to be neat and precise or the watch face would be ruined. To get the fine point they needed on their brushes, many of the women would twirl the bristles between their lips before dipping the brush in the radium paint again. Managers reportedly encouraged the practice by telling employees that radium would curl their hair and make their skin more beautiful.

Of course, it did nothing of the kind. Rather, continually ingesting small amounts of radium from the brushes led to an epidemic of cancer among the women, mostly in the bones of the head and jaw, but also many breast cancers. As it was later discovered, radium, because of its chemical similarity to calcium, accumulates on and in bones, bombarding them with alpha radiation. Many of the dial painters died from the disease, yet the companies were slow in correcting conditions at their

factories even after the source of the illnesses was identified.

Again, the working conditions of radium dial painters isn't analogous to any homeowner's exposure to radon. For one thing, the women working in the factories
probably ingested huge amounts of the radioactive paint.
According to one account, when the the bones of a dial
painter who died of cancer were later exhumed for
study, they contained so much radium that they showed
up on unexposed photographic film. In other words, the
film was not exposed to an outside source of X rays—
just to the bones. When the film was developed, the
bones were visible.

What concerns some scientists is that radon, a decay
product of radium, itself produces a radioactive daughter product, lead-210, that is a bone-seeking substance
(some of the cancers of the head and sinuses in radium
dial painters have been attributed to radon). The radioactivity of a substance is directly proportional to its half-
life: The longer it takes to decay, the smaller the amount
of radiation it's giving off. Lead-206 has a half-life of just
over 4.5 months, which means it releases its radiation
slowly and the radioactive dose to a given bone will be
low, even for a relatively large amount of radon. And
unless bombarded with large doses, the skeletal system
is quite resistant to radioactivity.

But once again, there has been very little research on
what role breathing or drinking radon might play in bone
cancer, principally in areas near uranium milling opera-
tions where radium levels in the water are sometimes
very high.

OTHER POSSIBILITIES

As we've seen, alpha radiation smashing through chromosomes can cause irreparable damage. After exposure
to radioactivity, chromosome aberrations show up in

many parts of the body, including the internal organs, bone marrow, and blood. When a biophysicist from the University of Salzburg took blood samples from 120 people working in an underground mineral spa in an area of Austria with very high natural radon levels, she found that at lower concentrations, radon produced proportionately more chromosome abnormalities in the blood than at higher levels.

"No one has ever really looked for things like chromosome damage to people living in houses with radon," says Dr. Harley. "Right now, chromosome damage in the blood doesn't mean much, because no one has been able to tie that to any health effects."

It's also interesting to note that, in addition to lung cancer, uranium miners showed a rate of noncancer respiratory disease more than three times higher than that of the general population. The statistics revealed an increasing number of deaths from obstructive lung diseases, such as fibrosis and emphysema, the higher the radon levels in the mine were. Writing in *The Annals of the New York Academy of Sciences*, the researchers theorized that the ailments may have been caused by diffuse radiation damage to the body of the lung.

QUESTIONS REMAIN

Obviously, there remain many more questions than answers about the health effects of radon. The answers will come—in time. But they can't come too soon for the thousands of people across the country who have tested their houses and discovered a lot of radon there.

Or for the radon-testing contractor who, after spending nearly an hour crawling around in the unvented crawl space of a house in New Jersey, found out later that the radon numbers under the house were among the

highest anyone had ever seen. "Several doctors have told
me that they *think* it didn't do me much harm," the contractor says now. "I want to believe them, but no one's
been able to say for sure if one bad exposure will lead to
health problems."

Or for Stanley and Diane Watras, who wait to see if
living in the most radioactive house in America will make
them sick someday.

CHAPTER
5

WHAT ARE
THE RISKS?

We live with risks every day, whether we choose to think about them or not. There's a statistical risk assigned to your walk to the grocery store, and to eating the steak you brought home and grilled on the barbecue. Some spoilsport has even calculated the risks of going fishing.

There's really not much sense in worrying too much about the danger in our day-to-day lives. Most risks we face are beyond our control. If a lovesick whale sinks your boat in the middle of the ocean, your time was probably up. Fretting about the possibility of an accident happening just leads to stress, which adds another risk to your life.

Of course, there are some risks we can control, and many we choose to take. Smoking is a voluntary, if needless, risk. So is picking a career as a fireman or spending your weekends rock climbing. If you want to live dangerously once in a while that's your privilege, but you should at least have an idea of what you're up against.

As long as you understand that there's a 1-in-500 chance your parachute won't open when you jump out of a plane at 7,000 feet, you know the score. After that, it becomes a *calculated* risk.

Radon poses an understandable, and controllable, risk to people who live around it. Health officials have attached lung cancer risk estimates to radon gas exposure at various levels. That way, the people who live in contaminated houses can have a clearer idea of the possible health effects they face. Only then can they make informed decisions about what to do about the problem.

Some choose to do nothing. After one radon survey in Maine, the majority of people who were told that their houses contained unsafe amounts of the gas chose not to do anything about it, at least not immediately. The agencies that conducted the tests told each and every person what the risks of inaction would be, but the homeowners may not have understood the magnitude of the risks.

Others use risk estimates to make choices on how much to spend fixing a radon problem. While most contaminations can be reduced to *acceptable* levels for $1,000 or less, it gets increasingly expensive to lower the radon level beyond that. And no amount of money will bring the reading to zero. As difficult as it is to put a dollar value on risks, many homeowners have been doing something like that by using dose estimates to decide how much work *needs* to be done to make their house reasonably safe. In other words, they're deciding how much radon risk they can comfortably live with.

THE LANGUAGE OF RADON

In order to understand radon risk estimates, you've got to learn a little of the language that's used to describe radon levels. No matter who tests your house for radon, or how they test it, the readings will be described in one of two ways: *picocuries per liter* (pCi/l), which refers to levels of radon gas in the house, and *working levels* (WL), a measure of exposure to radon's radioactive daughter products.

One curie equals 37 billion radioactive decays per second. A picocurie is 1 trillionth of a curie. One picocurie per liter (1 pCi/l) of air figures out to about two radon atoms per minute disintegrating in every quart of air in a room.

If that doesn't mean a lot to you, don't worry. Few people outside of a physics lab really understand pCi/l, or need to. As a homeowner who may want to test your house for radon contamination, it's a relative term, allowing you to compare your house to others and to judge the risks of living with the radon there. For example, Richard Guimond, the director of criteria and standards for the Environmental Protection Agency's (EPA) Office of Radiation Programs, says that, as a rule of thumb, living in a house that contains 10 pCi/l carries about the same lung cancer risk as smoking one pack of cigarettes a day.

Although pCi/l are easier to measure, radon gas isn't really the problem. It's exposure to the alpha radiation from radon's daughter products that is dangerous, and that is measured in working levels. As you might have guessed, the workers the term refers to are uranium miners. Working levels was a concept developed in the 1950s as a way of calculating the amount of radon daughters miners were, and are still, being exposed to underground. Mine safety standards are now regulated using working levels of exposure. They've since been adopted as a yardstick to estimate the risks of living with radon in houses.

Working levels are measurements of energy release. Officially, 1 WL is the amount of radon daughters the decay of which will result in the emission of 1.3 billion volts of electron energy. That sounds like a lot, but electron energy is more like static electricity than the type that powers a light bulb. A billion volts of electron energy isn't a whole lot, really. If converted to heat, it might raise the temperature of a cup of water a half a degree, or so.

In a typical house, 200 pCi/l of radon gas in the air convert to 1 WL. So, if the test reading for a house comes in at 4 pCi/l, the occupants are living with 0.02 WL of radon daughters.

But working levels alone aren't enough to estimate the risks of radon exposure. A working level is a measurement of decaying radon daughters, but risks are figured on the length of time you're exposed to that radioactivity. That cumulative exposure is expressed in working level months (WLM). Once more, it's based on the uranium miner's experience: 1 WLM is 170 hours (the amount of time the average worker spends in a mine over the course of a month) exposed to 1 WL.

Before 1960, uranium miners were routinely exposed to as much as 20 WL of radon. Under current occupational safety regulations, a person can't be exposed to more than 0.3 WL while on the job. A worker who spends a year in a mine at that level would accumulate 3.6 WLM of exposure (12 months × 0.3 WL). Long-term studies of uranium miners in the United States (both smokers and nonsmokers), have shown lung cancer effects at lifetime exposures as low as 40 WLM, about 12 years on the job for today's miners, but as little as 2 years for miners working in the 1950s.

But again, most of us aren't miners and we don't live underground, so there have to be some adjustments made to calculate risks to the rest of us who come in contact with radon.

For one thing, compared to a person living in a badly contaminated house, a miner has it easy. Workers only have to spend 8 hours or less in the uranium mine every

Radon Compared to Other Causes of Lung Cancer

pCi/l	WL	Comparable Exposure Levels		Comparable Risk*
200	1	1,000 times average ◀ outdoor level		▶ More than 75 times nonsmoker risk of dying from lung cancer
				▶ 4-pack-a-day smoker
100	0.5	100 times ◀ average indoor level		▶ 10,000 chest X rays per year
40	0.2			▶ 30 times nonsmoker risk of dying from lung cancer
20	0.1	100 times average ◀ outdoor level		▶ 2-pack-a-day smoker
10	0.05			▶ 1-pack-a-day smoker
4	0.02	10 times ◀ average indoor level		▶ 3 times nonsmoker risk of dying from lung cancer
2	0.01	10 times ◀ average outdoor level		▶ 200 chest X rays per year
				▶ Nonsmoker risk of dying from lung cancer
1	0.005	Average ◀ indoor level		
0.2	0.001	Average ◀ outdoor level		

*Based on lifetime exposure.

This shows how the lung cancer risks of radon exposure compare to other causes of the disease. For example, breathing 20 picocuries per liter (pCi/l) poses about the same lung cancer risk as smoking two packs of cigarettes a day. (Source: United States Environmental Protection Agency.)

day and usually work no more than five days a week. But
for many homeowners it's exactly the opposite: They
spend 8 or 10 hours a day *away* from the house and most
of the rest of their time exposed to indoor radon. Children and older persons may even spend 24 hours a day in
the house, particularly in the winter, which, ironically, is
when radon levels are at their highest.

So the numbers change significantly when you're
dealing with houses instead of mines. A WLM for a homeowner might be 340 hours (about 12 hours per day),
double the number for a miner. So, every year the homeowner exposed to the mine standard of 0.3 WL would
accumulate 7.2 WLM of exposure (24 × 0.3). That would
undoubtedly put homeowners up against an
unacceptably high risk.

How much risk is too much for homeowners? At
what radon level should they take action to fix a problem
in their house? *There is presently no federal standard
for radon, nor any nationally accepted safety level, so it
depends on whom you ask.* The American Society of
Heating, Refrigerating and Air-Conditioning Engineers
(ASHRAE) has proposed a guideline of 0.01 WL (2 pCi/l).
The Bonneville Power Administration (BPA), which was
part of the United States Department of Energy, established an action level of 0.025 WL (5 pCi/l). In Sweden,
by law, you must take remedial action within two years
if your house measures 0.27 WL (14.5 pCi/l).

It is hard to say what standard, if any, will be widely
adopted in the United States. Many experts are betting
that it will be the EPA's recommended level for remedial
action of 0.02 WL or 4 pCi/l in the living area of the
house (which may include the basement if there are
rooms there that are used extensively).

But even EPA officials are quick to point out that
they don't consider 0.02 WL "safe" or even "acceptable,"
only a realistic target to aim for. "It's not a particularly
low risk. We recognize that exposures to this level do

WHOM DO YOU TRUST?

How much radon in a house is too much? It depends upon whom you ask. Here are the "action levels," or levels at which various government agencies and private groups recommend doing remedial work to lower indoor radon:

Organization/Country	Action Level (pCi/l)
Sweden (for existing buildings)	11
Health and Welfare Canada	10
National Council on Radiation Protection	8
Florida Environmental Protection Agency	6
Sweden (for buildings undergoing renovation)	5
United States Department of Energy (DOE)	5
Union of Concerned Scientists	5
United States Environmental Protection Agency (EPA)	4
Sweden (for new construction)	2
American Society of Heating, Refrigerating and Air-Conditioning Engineers (ASHRAE)	2

present some risk of developing lung cancer," says the EPA's Richard Guimond. "However, it's probably financially impractical to try to get houses much below that."

Part of the EPA's dilemma is that for other carcinogens, such as toxic chemicals, pesticides, and air and water pollutants, the agency considers a risk of more than 1 death in a million people exposed to be too high. But by its own estimate, the average indoor level of radon in the United States, about 0.006 WL (1.3 pCi/l), causes 4,000 or more deaths per million. Obviously the old rules can't be applied to radon.

COMPARED TO WHAT?

Living in an environment that contains about 0.0005 picocuries per liter (pCi/l) of radon for a lifetime carries a risk of one death from lung cancer for every 100,000 people exposed. Below are some activities that pose the same risk.

Activity	Cause of Death
Being a 60-year-old man for 3 hours	All causes
Rock climbing for 15 minutes	Accident
Smoking 10 to 30 cigarettes (in a lifetime)	Heart/lung disease
Traveling 600 miles by car	Accident
Traveling 7,000 miles by air	Accident
Working for 30 hours in a coal mine	Accident
Working for 15 weeks in a typical factory	Accident

Source: Bonneville Power Administration.

What would Guimond do if *his* house tested out to 0.02 WL? "I'd certainly do something," he replies. "I'd take a look and see if I could do some easy things like sealing cracks in the basement to fix the problem. But if I couldn't eliminate the radon, I certainly wouldn't move out of my house."

In the end, it all boils down to you asking yourself and your family about how much risk you're willing to live with. In a way, "safe" is a relative term. We're never completely safe, and some people are more comfortable living with a higher degree of risk than others are. "The individual homeowner will have to be aware of the alternatives and then decide where to put his money, what level of cost and risk are acceptable," William J. Schull, a University of Texas radiation expert who helped the EPA arrive at its standards, told the *New York Times.* "I'd investigate my alternatives, then ask myself whether I'm prepared to entertain this risk or whether corrective measures are economically feasible. This is a decision a fair number of people are going to have to confront."

WAYS TO LOOK AT THE RISKS

So, after you've tested your home and read about the cost and effectiveness of various radon remedies in chapter 7, you'll have to come to your own decision. But to help you make it, the following statistics and estimates may move you toward a better understanding of just what those risks are and what they mean:

Living in a house at the EPA-recommended maximum of 0.02 WL for 70 years poses the same lung cancer

(continued on page 94)

Radon Compared to Everyday Risks

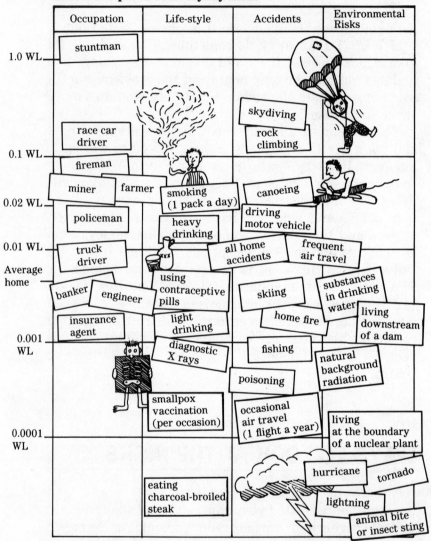

	Occupation	Life-style	Accidents	Environmental Risks
1.0 WL	stuntman			
	race car driver		skydiving / rock climbing	
0.1 WL	fireman			
0.02 WL	miner / farmer	smoking (1 pack a day)	canoeing	
	policeman	heavy drinking	driving motor vehicle	
0.01 WL	truck driver		all home accidents	frequent air travel
Average home	banker / engineer	using contraceptive pills	skiing	substances in drinking water
	insurance agent	light drinking	home fire	living downstream of a dam
0.001 WL		diagnostic X rays	fishing	natural background radiation
			poisoning	
		smallpox vaccination (per occasion)	occasional air travel (1 flight a year)	living at the boundary of a nuclear plant
0.0001 WL				
		eating charcoal-broiled steak		hurricane / tornado / lightning / animal bite or insect sting

How does danger posed by living in a radon-contaminated house compare with other everyday risks? Well, if your house is at the Environmental Protection Agency's (EPA) recommended "remedial action level" of 0.02 working level (WL), living there is safer than skydiving or driving a race car, but more dangerous than taking a skiing or fishing trip. (Source: Electric Power Research Institute.)

CHILDREN MAY BE
AT GREATER RISK

Young children living in homes with high radon levels are apparently at greater risk than their parents. Why? Partly because children are more active than adults, so they breathe more. Kids also spend a lot of time in the house, especially during the colder months. And, their lungs are smaller. So, while everyone in the house is exposed to about the same *amount* of radon in the air, children receive a larger internal *dose*. According to calculations presented in a recent Department of Energy report on radon, a ten-year-old child receives a radiation dose 30 percent higher than an adult male exposed to the same amount of radon.

Lung cancer is almost unheard of among children, and among people under age 40, for that matter. But growing evidence indicates that the younger a person is when exposed to radiation, the greater the risks of cancer later in life. "One must guard against a serious error of thinking, namely that children suffer the consequences of radiation-induced cancer less than do adults. *Exactly the opposite is true,*"writes Dr. John W. Gofman, in *Radiation and Human Health*. Although not writing specifically about the effects of radon exposure, Gofman contends that the number of cancers produced later in life for children under age 5 exposed to radiation is more than 3.5 times higher than for those exposed to the same amount of radiation between the ages of 16 and 20.

As yet, the EPA hasn't addressed the specific issue of children and radon, and has issued no special guidelines for homes with young children, but research on the subject continues.

risk as smoking about half a pack of cigarettes daily. It poses a cancer risk of about 1 in 65, approximately the same odds as drawing the ace of spades from a deck of cards on the first cut.

According to government calculations, if 100 people spent 70 years each living in houses with 0.02 WL of radon, about 3 of them would die of lung cancer as a result of the radon. (This is in addition to deaths attributed to other causes.) At 0.1 WL (20 pCi/l), an average of 13 would die of lung cancer. And at 1 WL (200 pCi/l), as many as 77 radon-induced lung cancer deaths could be expected in those 100 people.

If those same 100 people each lived in a 1 WL house for only 10 years instead of 70, between 14 and 42 of them would probably contract lung cancer as a result.

The EPA estimates that at 0.02 WL, a person faces a cancer risk equivalent to receiving more than 200 chest X rays per year. At about 0.3 WL (60 pCi/l), you might as well be getting 20,000 chest X rays yearly. Stanley Watras was told that living in his house with the 1984 radon levels there (14 WL, or 2,800 pCi/l) was as dangerous as receiving 455,000 chest X rays per year!

If you're having a backyard barbecue, the cancer danger from eating the charcoal-broiled steak is greater than from breathing the radon in the outdoor air (0.1 to 0.15 pCi/l, on average).

In terms of life expectancy, exposure to 1 pCi/l for a lifetime could shorten an average life by 25 days. Fifteen pCi/l reduces life expectancy by one year. In comparison, smoking a pack of cigarettes a day will cut about six years from an average life, being poor reduces it by four years, being overweight reduces it by two years, and breathing polluted air reduces it by 20 days.

Edward Radford, of the Radiation Effects Research Foundation in Hiroshima, Japan, has said that exposure to 0.015 WL (3 pCi/l) "for most of one's life" would increase the risk of lung cancer in nonsmokers by 75 percent or more and by about 15 percent in smokers (who already face higher risks).

The *lowest* total exposure that caused a significant increase in lung cancers among uranium miners was 60 WLM, or 1 WL for 10,200 hours. No study has shown a

LIFETIME RISKS

These are the lung cancer risks per working level month (WLM) of radon based on the length of time a person is exposed. One WLM equals exposure to 1 WL for 170 hours. One WL is equal to 200 pCi/l, an extremely high level of radon not frequently found in homes.

Exposure Duration	Lung Cancers*
1 year	13
5 years	66
10 years	130
30 years	380
Life	560

Source: National Council on Radiation Protection.
*Per 100,000 people

statistically significant cancer rate below that, so all cancer risk estimates for lower exposure are estimated downward from higher ones.

Breathing 0.005 pCi/l (0.0000009 WL) for life is about as hazardous as spending 15 minutes rock climbing or taking a 600-mile road trip. All carry a 1 in 100,000 risk of death, although obviously from different causes.

A 0.02 WL radon level translates into a whole body radiation exposure 100 times greater than federal guidelines allow for exposures at the boundary of nuclear power plants.

The Pennsylvania Department of Health says a person drinking water containing 20,000 pCi/l of radon (very high, but not unheard of, particularly in New England states such as Maine, where readings of 700,000 pCi/l and higher have been recorded) in a lifetime would have a 0.2 percent chance of dying of stomach cancer. Researchers at the University of Maine calculate the risk of death from drinking radon-tainted water for 60 years as 1 percent for every 100,000 pCi/l.

OTHER VOICES

Then there are people who accuse the EPA of scaring people with warnings, risk estimates, and cancer projections. "Some people think we've already gone overboard," Guimond told the Allentown, Pennsylvania *Morning Call.* "They would like us to have everything in such low-key wording that it's kind of ho hum. Well, you can't motivate anybody when you're that ho hum. You're trying to find a very fine line, where you have raised

LUNG CANCER RISKS

The following is a comparison of the lung cancer risks of living with various working levels (WL) of radon, and the reduction percentage that would be necessary to bring each level down to the EPA's recommended maximum of 0.02 WL.

Concentration (WL and pCi/l)	Risk of Death from Lung Cancer	% Reduction†
10.0 (2,000)	More than 75 times normal*	99.8
1.0 (200)	75 times normal	98
0.2 (40)	30 times normal	90
0.1 (20)	15 times normal	80
0.02 (4)	3 times normal	0

Source: United States Environmental Protection Agency.
"Normal" is the national average for lung cancer among nonsmokers. Risks are probably higher for smokers.
†Needed to reach 0.02 WL (4pCi/l)

enough consciousness about the health risks that this particular family is going to look into the problem but is not going to feel doomed. I'm not sure we've found that delicate balance yet, but we're trying."

Aside from the language the agency uses, there's a lively ongoing debate over whether the EPA's risk estimates are in the right ball park. A small but adamant minority of scientists feels that the numbers extrapolated from studies of uranium miners suggest far higher risks than actually exist or that they are skewed to misrepresent the danger to certain segments of the population.

One critic, Edward Martell, of the National Center for Atmospheric Research, believes the EPA overestimates the lung cancer threat for nonsmokers and underestimates it for smokers and those living in the house with them. Martell, who has been investigating synergistic effects of smoking and radon since the early 1970s, says, "When you burn cigarettes, radon decay products accumulate on smoke particles in room air and they're also passed through the cigarette into mainstream smoke. The two together are what really cause many of the lung cancers in our society.

"Lung cancer is very rare among nonsmokers," he continues. "If you're not a smoker and no one in your house smokes, lung cancer is going to be a rare disease of old age. I think the EPA risk estimates should be increased by 50 percent for smokers and reduced by a factor of 6 for people who don't smoke."

Others fear that growing up with atomic bombs, Three Mile Islands, and Chernobyls has created a national radiation phobia that has blown the radon issue out of proportion. In 1986, Nobel laureate Dr. Rosalyn Yalow expressed just such an opinion at a symposium sponsored by the Center for the Advancement of Radiation Education and Research. She complained that the publicity surrounding the health effects of radon has created a "mass hysteria."

The nuclear physicist also called the EPA's estimates of total yearly lung cancers from radon "clearly an exaggeration" and faulted the agency for extrapolating risks from uranium miners when "there are other things in those mines, like metals and dust and all sorts of stuff like that." She urged the EPA and other public health groups to "reexamine the basis on which they are giving these numbers that could end up costing the country tens and hundreds of millions of dollars unnecessarily to clean up something that is probably not nearly as hazardous as they think it is."

When asked about Yalow's comments by the *Washington Post*, Dr. Jacob Fabrikant, chairman of the National Academy of Sciences Committee on the Biological Effects of Ionizing Radiation, said, "It is clear that we have distorted the perspective on the issue of radiation . . . .We were born in a radioactive environment, developed in it, breathing it, and we've figured out a way to live with radon or we wouldn't be the way we are."

Thousands of hours of additional research, are needed before anyone can make the definitive statement on the health risks of living with radon. EPA officials, like environmental scientist John Davidson, freely acknowledge that predicting the dangers of living with radon in homes using evidence collected from studies of deaths among men working in mines is "a leap of faith, because there're no concrete health data on exposure at lower levels of radon you'll find in a lot of houses."

But, he adds, "A large portion of this country's yearly lung cancers can't be attributed to smoking, and we think radon is responsible for more of those than any other environmental factor. The EPA considers radon to be a very serious health issue. People who have significant amounts of radon in their homes and don't do anything about it are doing themselves a disservice."

CHAPTER

6

TESTING YOUR HOUSE

The results of the Environmental Protection Agency's (EPA) national radon survey won't be available until the 1990s. At this point, all there is to go on are test results from about 100,000 homes in the United States, far less than 1 percent of the total number of households in this country. And since it's been proven that radon levels can vary widely in houses on the same street, it's unlikely that anyone will be going too far out on a limb to predict exactly where radon problem areas are anytime in the foreseeable future.

So what can you do to set your mind at ease about whether you're living with radon as an uninvited guest? There's only one way: Test your home for it. In as little as a week, or within several months at most, you'll have a reasonably reliable idea about whether there are unsafe levels of radon in your house, and if so, how this gas is getting in.

Should *everyone* in the country test his or her house? Probably not. Is there any way to know which homes should be tested and which don't need to be? Probably not. Although the great majority of houses tested will contain only minute amounts of radon, at the moment it's impossible to guess which are the 72 million that are safe and which are the 2 million that aren't. If you want to find out which group *your* house belongs in, you'll have to find out for yourself.

"I think everyone who lives in a one- or two-family house should take a chance and test it, especially if the house has a basement or uses well water," says Andreas George, of the Department of Energy's Environmental Measurements Laboratory. "If I lived on the third floor or above of an apartment building, I probably wouldn't bother," she says.

Look at it this way: If someone told you that there was at least a 10 percent chance that termites were chomping away at your house, you'd probably have an inspector come over to check it out. The $40 or so it would cost might seem like a reasonable price to pay for peace of mind that the walls won't come crashing down around your ears someday. Well, the federal government is telling you that there's a one-in-ten chance that your house is contaminated with unhealthy amounts of radon. Others say the odds are even higher. For less than the cost of a termite inspection—as little as $10—you can run a simple do-it-yourself test that will tell you whether you need to give radon another thought.

Of course, there are lots of people who are more than willing to run the tests for you—for a lot more money. Most of these entrepreneurs are honest people trying to make a dollar. Others, seeing a chance to make more than their share of dollars by preying on the public's fears, are perhaps not so honest. Last year, after a television station in a large northeastern city ran a series of news stories on radon, fluorescent orange signs began appearing on street corners proclaiming (not very grammatically), "URGENT . . . HOME DETECTED NOW! . . . RA-

DON GAS CAN KILL . . . CALL AND HAVE YOUR HOME
TESTED AND FIND WAYS TO RID GAS!!!''
        When someone from the local Office of Consumer
Affairs began investigating the advertisements, he found
that they were being placed by a man who had never
even heard the word "radon" until a few days before.
After seeing the news broadcast, he sent away for some
$22 do-it-yourself detectors and was reselling them to
homeowners for $100. This was not illegal, but it wasn't
very helpful to frightened homeowners either.
        In 1986, 20,000 residents of the Reading Prong re-
ceived a five-page brochure in the mail that carried this
warning in boldfaced type: "OFFICIAL WARNING:

TESTING THE WATERS

The radon in your air could be coming from the radon
in your water. If your home water supply comes from a
well and your home's air tests high in radon, you ought
to be suspicious. If you live in one of the states where
the water is known to be naturally high in radon (see
chapter 2), you have cause to be even more suspicious.
        Suspicions are one thing, but you ought to be *sure*
before you spend Dollar One on remedial action. Radon
seeping in through the basement and radon coming
into the house via the water are two different problems, with very different solutions.
        Since a lot of the radon in water ends up in the air,
confirmation may come from selective air testing. If a
grab sample or an alpha-track detector placed in the
bathroom shows readings higher than in other parts of
the house, you've got another clue to work with (try to
keep the bathroom door closed as much as possible
during the test period).

YOUR HOUSE IS IN THE DANGER ZONE. YOU AND YOUR FAMILY ARE AT RISK." The booklet went on to warn that "Every breath you take in the 'safety' of your own home may be deadly."

The radon-testing firm that sent this comforting little message has since gone under, but not before needlessly scaring the daylights out of many people with its inaccurate "warning."

EPA spokesman Christian Rice says there have also been reports of "fly-by-night" operators who have tried to convince homeowners that radon testing can be done with a Geiger counter, which is not sensitive enough to detect most radon problems. He says there were also

You can also test the water directly. Terradex Corporation makes a special waterproof alpha-track detector that's enclosed in a weighted plastic housing. You drop the $30 tester into your toilet tank or well and leave it there for one to three months. You will get an answer within a month after sending back the kit.

Some private and public testing labs will analyze a sample of your water for radon, usually for less than $50. You'll be sent a special sample bottle to fill with water and mail back. When the bottles are returned, technicians will inject the sample with a special tolvene-based oil that makes it easier to measure the water's radioactivity.

For more information on water-testing programs in your state, contact your state environmental or public health agency (see Appendix A). Or, write to the following: Dr. Edward Vitz, Department of Physical Sciences, Kutztown University, Kutztown, PA 19530.

people going around selling mayonnaise jars covered with paper as "radon filters."

According to one newspaper account, another company was gathering "air samples" in plastic foam cups and then dropping two foam balls into the cups. Normal static electricity made the balls float, but the unscrupulous con artists told homeowners that the rate at which the balls fell illustrated the amount of radon in the house.

Be wary of anyone who's going door-to-door selling radon measurement services, or any company that attempts to sell you testing services over the phone or through frightening literature and official-sounding "statistics." Most legitimate, trained testing contractors don't need to drum up business that way.

In any case, you'll probably never need a professional to test your house. If you can use a pencil and lick a stamp, you can do the job yourself and save a lot of money.

THE LONG AND SHORT OF TESTING

The EPA recommends a multi-stage testing process. If you "flunk" the first test by coming up with a high radon reading, you go on to stage two. If you "pass" the second stage, there's probably no need for further tests.

The first stage is the screening test, to initially determine if high concentrations of radon exist in your house. The EPA says these tests should be "inexpensive and relatively fast, so that many houses can be tested without wasting time and money on houses that don't pose a health threat." If the reading from this initial test is 0.02 working level (WL), which equals 4 picocuries per liter (pCi/l), or below, the agency says "follow-up measurements are not recommended, but are at the discretion of the resident."

If your results are between 0.02 and 0.1 WL (4 to 20 pCi/l), the EPA says that you should be aware that there are serious concerns about long-term exposure to that amount of radon, but that at these concentrations there's no large increase in risk if you live there without fixing the problem for another year. So, before you spend money to get rid of the radon, follow up with another, more long-term test.

Should you be one of the unlucky few whose house gets an initial reading of 0.1 to 1 WL (20 to 200 pCi/l), you don't have the luxury of so much time to run a long-term retest. The EPA strongly suggests that you get a retest, using a quick test method that will give you a confirmation of the first test within a week or two. Although a long-term retest will give you a more accurate picture of your average annual exposure to radon, the EPA doesn't advise using it exclusively "because an additional 12 months of exposure *could* cause a significant increase in health risk."

Finally, if your results show levels higher than 1 WL (200 pCi/l), call your state health department or regional EPA office (see Appendix A) *immediately* for advice on how to reduce the concentrations.

Where screening tests are usually done quickly, measuring radon levels for anywhere from a few minutes to a few weeks, follow-up tests are used to estimate exposures over a longer period of time—from several months to a year. Radon monitors left in several parts of the house for 12 months will show what the average radon concentrations were over a variety of weather and household conditions.

For example, radon levels are almost always higher during the winter, when the house is closed up. A screening test conducted in the colder months will give you an idea of the "worst case" conditions in your house. But radon levels can vary by a factor of 10 between winter

WHEN TO RETEST

An inexpensive do-it-yourself screening test is a good way to get a rough idea of the radon levels in your house. When the results are in from this initial monitoring, most people will find that their homes fall well within a safe range. But some homes will show radon concentrations that at least warrant further testing. The EPA's Office of Radiation Programs recommends these follow-up actions based on the results of the first screening test:

Screening Measurement Result	Recommended Action
Greater than 1 WL (220 pCi/l)	Perform a follow-up test as soon as possible; expose the new detectors for no more than 1 week, keeping doors and windows closed as much as possible during the testing
About 0.1 WL (20 pCi/l) to 1 WL	Perform a follow-up test, exposing detectors for no more than 3 months; Keep doors and windows closed during test
About 0.02 WL (4 pCi/l) to 0.1 WL	Retest, exposing the new detectors for 1 year or making 1 short-term measurement during each season of the year
Less than 0.02 WL	No follow-up testing is necessary

and summer. And since radon health risks are proportional to total exposure, not peak exposure, you'll want to know what the average levels were for the entire year.

Unless the radon levels from your first test are unusually high, in the 200 pCi/l range, you shouldn't spend any money trying to fix the "problem" until you've run a follow-up test to make sure there really is one. Remedial action to permanently lower the radon concentrations is probably going to cost at least a few hundred dollars, maybe more. You really can't make an informed decision about how much to spend and which way to spend it until you're sure there's a problem to begin with.

If for no other reason, it's smart to retest just to make sure that the first test kit wasn't defective and its reading inaccurate. Testers sold by companies that passed the EPA's protocols have been proven to be reliable, but not perfect. A certain number of false readings, high and low, are inevitable. Testing again to confirm high readings will protect you from mistakes. It's possible to get a bum detector once, but the chances of getting another from a separate order are infinitesimal (provided you're dealing with a reputable company).

WHERE AND WHEN TO TEST

Where in the house you place radon detectors and what the conditions inside and outside the house are during the test can greatly affect the accuracy of the results. There are ideal times and places in your home to test for radon.

IDEAL TIMES TO TEST

Ideally, the testing should be done in the late fall or early spring, times when the house is closed up and the heating system isn't turned on yet. There is less natural ventilation to dilute radon concentrations then. Next best are tests conducted during the winter months. The heating may throw the readings off slightly, especially if you have a forced-air heating system, but you should still get a fairly accurate reading.

It's understandable that you might not want to wait until the ideal times to run your screening test. But if you must test in the summer, at least try to approximate winter living conditions by keeping the house closed up as much as possible during the test period. You don't have to live like a hermit, but you should try to keep the ventilation rate as low as possible. Leave the windows closed if you can and open exterior doors for no more than a few minutes as you come and go. You shouldn't run large window or attic fans or air conditioners during the test.

If you'll be using a detector that measures for three days or less, start these "closed-house" procedures at least 12 hours before starting the test. And keep an eye on the weather reports. It's best not to test when a storm is predicted. High winds can create differences between the air pressure inside and outdoors, which will throw off the measurement. And the sudden drop in barometric pressure that often accompanies bad weather may draw more radon out of the soil and make your indoor levels look worse than they really are.

IDEAL PLACES TO PUT TESTERS

Place the tester in the lowest finished space in the house in a spot where it won't be disturbed. If there's a basement, put it there, preferably in an interior room with few or no windows and tight doors. Don't stick the tester in a cupboard, closet, or nook in the foundation, and keep it away from any possible sources of air movement such as heating vents, clothes dryers, fireplaces, or badly cracked exterior basement walls. To avoid the diluting effect of floor drafts, place the detector at least 20 inches off the floor.

Screening test procedures are designed to give you an indication of the *maximum* amount of radon getting into the house. Testing that way offers better insurance against false readings, but the results aren't meant to reflect real-life conditions. Unless someone is living in the room where the test was taken, the actual daily levels you and your family are exposed to are probably two or three times lower than the screening results. So, again, don't panic if the initial test comes back high. That only indicates that you need to do some more investigating.

For a follow-up test, you'll want to place at least two new detectors in different parts of the house and leave

ROOMS WITH RADON

Radon levels in the basement are almost always the highest in a house, usually at least twice as high as in other parts of the house. For instance, the following are average radon levels in various rooms of 85 houses sampled by researchers from the University of Maine:

Room	Radon (pCi/l)
Basement	6.4
Bathroom	3.4
Living room	3.1
Bedroom	3.0
Outdoor air	0.8

WHERE TO PUT DETECTORS

All houses are not created equal. The best places to put your radon detectors vary depending on the type of house you live in. Here are some options:

1. Your house has an unfinished basement used for storage and laundry. The ground floor has a kitchen and living room, and the second floor has two bedrooms. Conduct your screening test in the basement and place your follow-up detectors (if needed) in the living room and one or both of the bedrooms.

2. You have a split-level house built over a garage on the lower level and with a den and one bedroom built several feet higher than the garage. There's a kitchen on the middle level over the garage and more bedrooms on the level over the den. The screening measurement should be made in the lowest bedroom on the floor next to the garage. The follow-up measurements

them there for up to a year. There's no reason why you can't do these tests yourself. But because of the length of the monitoring, your choice of testers is pretty much limited to the *alpha-track* type described later in this chapter. These small plastic detectors cost $20 to $50 but are very accurate and can be left in place for a year or more without deteriorating. The only other do-it-yourself option is to run four short-term tests at three-month intervals and average their results. You won't save any money that way, but you'll at least get ongoing "interim reports" on the radon levels in the house.

Either way, make your measurements in living areas on at least two levels of the house. Choose the one room should be made in the same lower bedroom and one of the upper bedrooms.

3. You live in a farmhouse with a dirt-floor cellar. The ground floor contains a kitchen and a living room, and the bedrooms are on the upper floor. Put the screening tester in the living room and follow up with detectors in the same living room and one of the upper floor bedrooms.

4. You live in a multi-story apartment building with a finished basement containing a laundry. Take a screening measurement of the basement; if it's above 4 pCi/l, one measurement should be made in the bedroom and living room of each apartment on the next several floors up.

5. Your home is a one-story, slab-on-grade house (no basement). The screening measurement should be made in the bedroom and the follow-up in another bedroom and the living room.

where your family spends the most time and put one there. Place the other one in an upstairs bedroom. If your house has only one story, put detectors in two different rooms on that level, a bedroom and living room, for example.

Since children are at greater risk from radon than adults, put a detector in the bedroom of each child younger than 12 and make sure that any playrooms they use are tested separately, too.

Don't use the kitchen as one of your test rooms. The exhaust fan and particles generated by cooking will throw off the measurements. You shouldn't put testers in bathrooms, either. You don't spend much time there, anyway, and the humidity and fan will affect some detectors.

However, if you suspect that significant amounts of radon may be coming in via your well water, put a detector in the bathroom. If the radon levels in your water are elevated, it will show up as elevated airborne radon in the bathroom. You should at least be able to get a ball park figure from this type of testing.

The same technique can be used for other suspected radon inlets: If you have a sump pump, basement toilet, a badly cracked exterior basement wall, or other foundation penetrations, you can put a tester nearby. Or, you can put a tester in your crawl space to measure levels there. Just be sure to keep the results of those tests separate and don't include their measurements when you average the results of your other follow-up testers.

No matter where you put detectors in the house, it's *crucial* that you keep very careful records. Start a log book and write down the identification number of each detector, when and where it was placed, and how long it was there before you sent it back for analysis. The test results for a group of detectors will come back from the lab with the results listed by each tester's identification number. It's up to you to have a record of what was where.

ANNUAL AVERAGE RISK

When the results are in from your follow-up measurements, you'll be in a much better position to judge the actual amount of radon exposure you and your family are living with on a daily basis. When all the follow-up measurements from different parts of the house are in, you'll need to combine them together to come up with the annual average concentration. To do that, simply add all the results together (do not include the measurements from your screening test) and divide that total by the number of testers used.

For example, suppose you put detectors in the living room, a child's bedroom, and the master bedroom, and the results were 21 pCi/l, 13 pCi/l, and 9 pCi/l. The average would be 14 pCi/l or:

$$\frac{21 \text{ pCi/l} + 13 \text{ pCi/l} + 9 \text{ pCi/l}}{3} = 14.3 \text{ pCi/l}$$

That number is reality, or as close as you're probably going to get. Once it's in hand, you can begin to make further investigations about the sources of radon in your house, and what you want to do to eliminate them. Base your risk estimate on this average annual exposure and use it to decide how quickly you should proceed with remedial action.

YOUR TESTING OPTIONS

Before going any further, let's look at the various types of tests available to homeowners. The detectors themselves range from $10,000 high-tech machines that must be carefully calibrated and used only by trained professionals to simple bags of charcoal that cost about $10 and can be used by anyone.

Following are the testing options, both do-it-yourself and professional, available to you now. (For sources of test kits and names of manufacturers and testing contractors who have passed the EPA's proficiency tests, see Appendix B.)

DO-IT-YOURSELF KITS

The two types of do-it-yourself test kits currently available, *activated charcoal monitors* and *alpha-track detectors*, make a good team. The charcoal type will give you a fast reading for screening tests, and the alpha-

THE TYPES OF RADON DETECTORS AND WHAT THEY MEASURE

Detector Type	Unit of Measure
Activated charcoal monitor	(pCi/l)
Alpha-track detector	(pCi/l)
Continuous radon gas monitor	(pCi/l)
Continuous working level monitor	(WL)
Grab radon gas sampler	(pCi/l)
Grab working level sampler	(WL)
Radon progeny integrated sampler	(WL)

Source: United States Environmental Protection Agency.

track type can be left in place for up to a year for follow-up measurements.

ACTIVATED CHARCOAL MONITORS

Cost: $10 to $25
Testing duration: Three to seven days

Especially useful for screening tests, activated charcoal monitors are left in place for no more than a week and then returned to the manufacturer for analysis.

Charcoal monitors come in several forms. Some companies sell them as small metal cans covered with a fine-mesh screen and filled with 1 to 5 ounces of activated carbon (usually coconut charcoal). Other makers use slightly more charcoal and a treated paper bag instead of a can. Both seem to work equally well and are very accurate. They are capable of measuring radon concentrations as low as 0.03 pCi/l.

To use a charcoal monitor, first unscrew the can's cover or remove the paper bag from its foil pouch, depending on which type you're using (make sure they are tightly sealed when they come from the manufacturer and keep them that way until you're ready to deploy them). Then place the detector in an unobtrusive spot at least 20 inches off the floor and 4 inches from other objects. Mark the date, identification number, and location of the detector in a notebook and on the log sheet that comes with the test kit.

Nothing should interfere with the airflow around the detector. Also, avoid putting the detector in areas near high heat, such as by a fireplace, wood stove or furnace, or where the humidity is high. Because of the relatively short sampling time of these detectors, take special precautions to assure that the house is kept closed as much as possible while the detectors are in place.

In three days to a week (the manufacturer's instructions will tell you how long to leave this monitor in place; you can usually deviate by as much as six hours either way with no appreciable loss of accuracy), retrieve the detector and inspect it for signs of damage. Note any changes, then put it back into its bag, or reseal the can, record the date on the log sheet, and promptly return it to the manufacturer for analysis.

Charcoal detectors work by absorbing radon gas from the surrounding air. Back at the lab, technicians use a $6,000 to $10,000 machine called a *sodium iodine gamma scintillation detector* that counts the gamma rays emitted by the radon decay products on the charcoal.

ADVANTAGES OF CHARCOAL DETECTORS

- Quick results
- Inexpensive
- Can be sold and analyzed by mail
- Very easy to use
- Operate with no external power, so they can be placed anywhere in the house
- Accurate

DISADVANTAGES OF CHARCOAL DETECTORS

- Short testing period doesn't give a good picture of long-term levels of indoor radon (charcoal detectors will measure radon for no more than a week; if they're left in place longer, they'll still only give you a measurement for the previous seven days)
- Sensitive to temperature and humidity

ALPHA-TRACK DETECTORS

Cost: $20 to $50
Testing duration: One month to one year

Commonly called *track-etch detectors,* these round plastic cups are not much bigger than a twist-off bottle cap. They're useful for follow-up measurements of up to a year. Placed in selected locations around the house, two or three of these detectors can give you an overall picture of average yearly radon exposure through different seasons and house conditions, for a cost of $100 or less. (Individual detectors usually cost about $50 but are often considerably cheaper when you order several at a time.)

Alpha-track detectors are usually mailed from the testing company in sealed aluminized plastic bags. To deploy them, you cut open the package (save it, you'll need it to send the detector back for analysis), remove the detector cup, and unspool the paper tape wrapped around it. Do not remove the plastic cover of the cup.

Usually the detector will begin recording radon data as soon as the bag is opened, although some brands have a seal that covers the holes on top that must be peeled off first. Place the monitor upright (holes pointing up) on a shelf, bookcase, table, or any flat surface. The paper tape spooled around it can be used to hang the tester on a wall or from a shelf or basement ceiling beam with a thumbtack. The strip may also have printed spaces for logging in your name, the date, the location of tester placement, and its serial number. To be safe, keep a separate record of those data in a notebook of your own.

It doesn't much matter where in a room you put these detectors, as long as the spot is free of drafts and out of the way enough that the cup won't be disturbed by normal activities. But the shorter the testing period, the more you should try to maintain a reasonably closed house. If the detector will be in place for a year, you don't have to worry. If you're testing for three months, opening the windows for a few days during that period won't invalidate the test results. But for a one-month test, the results could be affected by too much ventilation. And if you are planning major remodeling projects or replacing a furnace or central air conditioning unit, consider delaying your monitoring program until the work is completed.

At the end of the test period, fill in the removal date on the paper tape, put the detector back in its aluminum bag, fold the open end several times, and seal with a paper clip or tape, and then mail the kit back to the maker for analyzing as soon as possible. (If you've misplaced the original bag, wrap the detector in several layers of aluminum foil and tape it closed.)

Inside the plastic detector cup is a sheet of treated plastic measuring less than 1 inch square. When alpha particles emitted by the decay of radon gas strike the plastic, they etch minute "tracks" on its surface. When the detector is returned to the laboratory, the piece of plastic is placed in a caustic solution that accentuates the tracks, which are then counted manually or automatically under a microscope. From the number of tracks, the laboratory can determine the radon concentration in the house where the detector was placed.

Alpha-track detectors have been in use for a number of years and have generally proven themselves to be reliable. The detectors made by one company, the Terradex Corporation (see Appendix B), have been used to test more than 50,000 houses in the United States and were chosen by the Commonwealth of Pennsylvania when it instituted a program to give free radon monitors to homeowners in the Reading Prong.

However, the EPA cautions that the detectors aren't infallible. The results, they say, can show a high degree of variability, particularly at low radon concentrations and if the laboratory doing the analysis inspects only a small section of the plastic detector for alpha tracks. Therefore, it seems very important that you buy an alpha-track tester only from a company that has passed the EPA's screening program, and when possible, to use more than one detector in the house as a fail-safe measure.

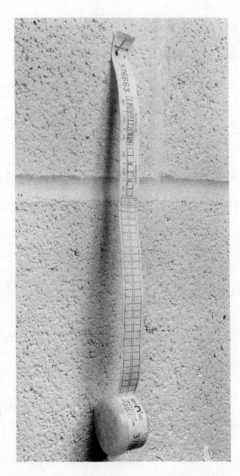

An alpha-track radon detector hangs on this basement wall. This type of tester, which costs $50 or less, can be left in place to measure average radon levels for a few months up to an entire year. (Photo by Mitchell T. Mandel.)

ADVANTAGES
OF ALPHA-TRACK DETECTORS

- Not temperature sensitive (no low-temperature limit; temperatures up to 160°F acceptable)
- Resistant to humidity
- Can measure for up to 12 months
- Need no external power source
- Can be sold and analyzed by mail
- Easy to use
- Usually accurate

DISADVANTAGES
OF ALPHA-TRACK DETECTORS

- Long measurement time necessary (not accurate for test durations shorter than one month)
- Large inherent variability

HOME RADON DETECTION DEVICES
(IN DEVELOPMENT)

Cost: $150 (projected)
Testing duration: Continuous

At the time this book went to press, the Sun Nuclear Corporation (415-C Pineda Court, Melbourne, FL 32940) had received an EPA grant of several hundred thousand dollars to develop and manufacture a radon monitor that will continually monitor radon levels in a house and warn occupants if the concentrations reach unsafe levels. Thomas L. Powers, the company's president, says the battery-operated device will "appear and function similarly to a smoke detector and will be small, lightweight, and unobtrusive. It will lend itself to placement on the piano in the living room or mounting to a floor joist in the crawl space."

Once in place, he adds, the monitor would begin automatically collecting and storing data on radon and radon daughter concentrations in the house and would provide a continuous display of the result. External outputs would be included so that the monitor could be rigged to turn on fan systems when radon levels reach a preset point, or to trigger remote alarms.

Based on Powers's description, such a device would seem useful both for conducting long-term radon tests and for continually monitoring radon levels in houses that have undergone remedial action. If the company sticks to its projected price of $150, using its monitor would be cheaper than running several alpha-track tests yearly. The product is expected to be available sometime in 1987.

PROFESSIONAL TESTS

It may be necessary, in some situations, to call in professionals to help you test a house for radon. While a combination of activated charcoal and alpha-track tests is certainly capable of yielding accurate short- and long-term radon readings, it's not unreasonable to get another opinion before undertaking expensive repairs.

Or, you may want immediate results from a test—sort of a Polaroid picture of your home's radon levels. There is equipment that can get you that information in less than an hour. You might want to use such a service to confirm a very high screening test you ran yourself, or to get a quick answer on your house if a neighbor's is found to be badly contaminated.

Prices for professional testing services can vary greatly. You can expect to pay at least $40 to get someone to come out to your house to run a quick test. Long-term sampling may cost more. It wouldn't be fair to quote

(continued on page 124)

FINDING OUT BEFORE YOU BUILD

Radon-proofing a new house as it's being constructed is almost always cheaper and less complicated than having to go back and do the job once the house is completed. If you knew there was a source of radon in the piece of land you've chosen to build a house on, you could take steps to insure that the radon gas couldn't find its way into the house.

Unfortunately, there's no foolproof method for testing open land. Until there is, people planning to build will have to make do with the best tests available now. And at this point, that's a hit or miss proposition.

The problem is that the amount of radon coming out of the soil, known as the *radon flux*, can vary from place to place, even on a half-acre building lot. There's no practical way to test an *entire* piece of land, so you could miss a hot spot that would affect radon levels in your house after it's built. However, even if open land testing is no guarantee against radon, it's worth trying. And the more areas of your lot you test, the better your chances of getting an accurate picture of the situation.

There are several ways to test your land. One maker of charcoal radon testers, Airchek (see Appendix B), sells a special kit designed for that purpose. The kit, which sells for less than $20, includes detailed instructions, one of the company's bag-type activated charcoal testers, and a small fold-up cardboard "tent."

The company recommends using two kits to test the average building lot. Once you have decided on the approximate dimensions of your house, you divide the lot in half and clear a small circular patch of ground about 2 feet across down to bare ground in the middle of each half.

Then, you simply hang a tester from the center of the inside of each of the cardboard tents, place one over each bare spot, and bank soil around the bottom edges to keep air from blowing inside. After 16 to 24 hours, the test is completed and you can send the bags back to the company for analysis.

If you're planning a house with a basement, Airchek suggests placing each charcoal bag in a 4- to 8-inch-deep hole before covering it with the tent.

The Terradex Corporation has devised a somewhat different open land test and even developed a special version of its Track Etch detector for the purpose. Larger than their in-house tester, this one looks like a plastic drinking cup, with the plastic alpha-track detector attached to the bottom. Terradex's instructions call for digging a hole 15 inches deep at the corners of your proposed building perimeter and burying one of the $26 devices, open end down, in each hole. After three weeks, you retrieve the cups and return them to Terradex.

The results of these relatively crude tests can't possibly pinpoint what actual radon levels you would expect to find in the finished house, nor does anyone claim that they will. Use the readings as a rule of thumb to "guesstimate" the approximate radon concentrations you might find in the crawl space of the house, and work from there.

prices for services here, but I'll give you an idea of how much it costs the testing company to buy the equipment.

Again, the EPA has developed measurement protocols for each type of system, but it doesn't guarantee that every company will follow them, or keep its equipment perfectly calibrated. You are safest, then, sticking with a company that has passed the agency's proficiency program.

CONTINUOUS
RADON MONITORS (CRM)

Equipment cost: $2,500 to $10,000
Testing duration: At least 24 hours

A CRM is a sophisticated machine that samples air by pumping it through a special filter that traps dust and particles, then pulling the air into a chamber called a *scintillation cell* where the radon gas decays. The alpha particles that are produced strike the zinc-based coating of the cell and are detected by a special photo tube that generates electrical signals. The signals, which tell how much radon is in the air, are processed and either stored in the memory of the machine or printed out on computer paper. In addition to the monitor itself, the technician will need a readout device, printer, and an airflow-rate meter.

The CRM is mainly an investigative tool. While do-it-yourself detectors give you one reading of the total concentration over a given period, a CRM measures in "real time," showing the hourly peaks and dips in radon levels and how house conditions affect them, as well as total levels. It could also be used to pinpoint a problem with radon in your water, since levels should increase after the water is run for 10 or 15 minutes.

It takes several hours for this machine to stabilize its readings once it's brought into the house. For that reason a test period of at least 24 hours is recommended and the results of the first 2 hours should probably be disregarded. In general, the longer the test, the more certain you can be about the results. And the shorter the test, the more you'll want to control the indoor environment by keeping the house closed up for at least 12 hours before monitoring.

As long as the machine is calibrated and maintained properly, there's no great skill involved in setting it up in the house. The technician will know to place the air intake on a stool or table at least 20 inches off the floor and away from outside walls, windows, doors, and vents.

ADVANTAGES OF CRM

- Hourly results available on-site; no waiting for detectors to come back from a lab
- Relatively short measurement duration
- Can track radon highs and lows to help you get a clearer picture of what affects levels
- Accurate for tests of at least 24 hours

DISADVANTAGES OF CRM

- Equipment is expensive and must be carefully maintained and calibrated
- Testing must be done by trained operator
- Some models are heavy and awkward to move
- Higher degree of error with short test duration

RADON PROGENY INTEGRATING SAMPLING UNITS (RPISU)

Equipment cost: $500 to $3,000
Testing duration: Three to seven days

Originally developed to measure radioactivity in buildings suspected of being contaminated with uranium mill tailings, the RPISU measures radon decay products and produces readings that are expressed in working levels. In other words, it tells you the extent of your exposure not to radon gas but to the radioactivity from its decay products. And since it is these decay products that cause biological harm, the readings can help you assess the health risks. An RPISU can detect radon daughter concentrations as low as 0.0005 WL after a one-week sample.

Cheaper to purchase than continuous monitors, these machines collect samples by means of an air pump that pulls air through a detector assembly that includes a filter and something called a *thermoluminescent dosimeter* (TLD). Suffice it to say that the TLD measures the radiation emitted from radon decay products.

After the unit has been in place for three to seven days, it's removed and the detector assembly is detached and returned to the lab for analysis, which is done by measuring the light given off by the TLD when it's heated. After a reading is taken, the TLD can be cleaned and put back in the machine for reuse.

ADVANTAGES OF AN RPISU

- Measures radon decay products
- Results available from relatively short measurement times
- Has been in use for more than ten years

DISADVANTAGES OF AN RPISU

- Limited to use in house locations where electric power is available
- Machine must be installed and picked up by trained technician

- Readings may be affected by high levels of dust or cigarette smoke in house

GRAB SAMPLING

Equipment cost: $2,500 to $10,000
Testing duration: 2 to 10 minutes

As the name implies, this testing method involves "grabbing" a single sample of indoor air in a container and analyzing it. There are two types of grab samples available: one for radon gas and the other for radon decay products.

For radon gas testing, a sample of from 40 to 800 cubic inches of air is pumped into a special flask and the air is quickly brought to a laboratory where the radon gas concentrations are measured.

More widely used is the radon decay products grab sample. A sample of up to about 105 quarts of air is sucked through a 1- to 2-inch-diameter filter that traps anything larger than 0.8 microns, which includes radon daughters (the process takes 5 minutes or less). The filter can then be analyzed on the spot to measure the total alpha particle activity on its surface. Using an elaborate conversion table, the technician can then calculate the working levels of radioactivity in the air.

Grab samples have their uses. They get you an answer more quickly—within an hour in most situations—than other do-it-yourself testing methods. And the tester will know exactly what the weather conditions were during the sample. A grab sample is easy to take, inexpensive, and can be used to confirm high readings obtained through other test methods.

But the limitations of such a short-term test are obvious. The EPA puts it this way: "The uncertainties associated with grab sample results are far larger than for any

(continued on page 130)

THE STORY SO FAR

The following is a state-by-state breakdown of the results of 57,015 radon tests conducted during the past several years. The table shows the total number of houses tested in each state, the number of homes that showed readings at or above the EPA's "action level" of 4 pCi/l, and the percentage of houses that number represents. Obviously, Pennsylvania is the most-tested state in the United States, partly because the Commonwealth of Pennsylvania gave away free test kits to people living within the Reading Prong.

However, none of the numbers represent a random sample of homes in each state, so they can't be considered as representative of national radon levels overall. For example, most of the houses tested in New Jersey are probably located in the part of the Reading Prong that underlies some cities in that state. Many of the people who had their houses tested probably did so because they'd heard about the radon problem in their area. It would be a mistake, then, to assume that radon levels in the *entire* state are as high as these numbers would suggest. What the results do show is that there are areas of radon contamination in almost every state in the country.

State	Number of Homes Tested	Number of Homes Tested w/ Radon at 4+ pCi/l	% of Homes Tested w/ Radon at 4+ pCi/l
Alabama	11	1	9.1
Alaska	49	20	40.8
Arizona	106	27	25.5

State	Number of Homes Tested	Number of Homes Tested w/ Radon at 4+ pCi/l	% of Homes Tested w/ Radon at 4+ pCi/l
Arkansas	10	3	30.0
California	721	40	5.5
Colorado	289	141	48.8
Connecticut	170	45	26.5
Delaware	41	7	17.1
District of Columbia	46	8	17.4
Florida	229	36	15.7
Georgia	87	17	19.5
Hawaii	2	0	0.0
Idaho	709	293	41.3
Illinois	264	92	34.8
Indiana	210	73	34.8
Iowa	83	35	42.2
Kansas	27	5	18.5
Kentucky	107	32	29.9
Louisiana	26	1	3.8
Maine	1,535	395	25.7
Maryland	302	100	33.1
Massachusetts	266	69	25.9
Michigan	139	45	32.4
Minnesota	281	110	39.1
Mississippi	14	1	7.1
Missouri	84	12	14.3
Montana	303	102	33.7
Nebraska	12	4	33.3

(continued)

THE STORY SO FAR—*Continued*

State	Number of Homes Tested	Number of Homes Tested w/ Radon at 4+ pCi/l	% of Homes Tested w/ Radon at 4+ pCi/l
Nevada	312	44	14.1
New Hampshire	134	64	47.8
New Jersey	4,273	1,140	26.7
New Mexico	590	158	26.8
New York	4,916	662	13.5
North Carolina	107	28	26.2
North Dakota	50	15	30.0
Ohio	455	188	41.3
Oklahoma	19	3	15.8
Oregon	5,574	281	5.0
Pennsylvania	20,527	11,770	57.3
Rhode Island	19	8	42.1
South Carolina	33	1	3.0

other (test) methods." The reason, of course, is because of the extremely short sampling period. The results tell you only what the radon conditions were for a 5- or 10-minute period. It's entirely possible that if the test were conducted a few hours later, the results might be completely different—significantly higher or lower. Grab samples shouldn't be used as your only radon test, and

State	Number of Homes Tested	Number of Homes Tested w/ Radon at 4+ pCi/l	% of Homes Tested w/ Radon at 4+ pCi/l
South Dakota	608	173	28.5
Tennessee	515	148	28.7
Texas	475	40	8.4
Utah	38	6	15.8
Vermont	91	27	29.7
Virginia	494	99	20.0
Washington	11,269	1,510	13.4
West Virginia	16	4	25.0
Wisconsin	240	54	22.5
Wyoming	137	66	48.2
Total	57,015	18,203	31.9 (average)

Source: Results of homeowner-requested tests conducted by Terradex Corporation and Airchek.

the results of a single grab sample should *never* form the basis for decisions about remedial action.

Nor should you use a grab sample to determine whether a house you're negotiating to purchase is free from radon. In Pennsylvania, where concerns about radon have caused headaches among home buyers, sellers, and real estate agents, more than 1,000 houses were tested with one type of grab sample, the Kusnetz method, as part of the sales transactions. Then the Pennsylvania Department of Environmental Resources cried foul, saying the method of testing failed to meet federal environmental standards.

The problem, the EPA's Richard Guimond told a local newspaper, is that "Kusnetz is one of the easiest tests to fudge." All a home seller would have to do, he explained, was open some windows before the testing company arrived and it would look like the house was radon-free. And that's apparently what happened in Montana in 1983, when the United States Department of Housing and Urban Development began requiring radon tests for all federally financed houses, and the Kusnetz method was used for many of the tests. "In Butte," said Guimond, "they were opening their windows and airing their houses out like crazy."

The moral seems to be, use a grab sample when it suits you and when you're sure of the house conditions before *and* during the test. But don't bet on it.

ADVANTAGES OF GRAB SAMPLING

- Quick results
- Equipment is portable
- Many houses can be tested in one day
- Conditions during sampling will be known

DISADVANTAGES OF GRAB SAMPLING

- Very short measurement duration not a reliable indicator of long-term radon levels
- Homeowner must be especially careful to keep the house closed for at least 12 hours before, as well as during, the test
- May give conflicting readings, depending on the time of day the testing was done

THE NEXT STEP AFTER TESTING

As you can see, there are no shortages of ways to test your house for radon. The testing process isn't always quick, and sometimes that can lead to some stress. But it's still worth the wait to be sure. Again, keep in mind that the health problems associated with radon are almost all linked to exposure over decades. So unless your screening tests hint that you're dealing with a very large radon problem, there's no compelling reason not to wait a few months, even a year, to get the big picture.

Since I wrote my first story on radon more than five years ago, I've received countless calls and letters from panicked friends, relatives, and strangers whose initial tests came back high. But of all those people, *only three* that I know of still had reason to be concerned after follow-up testing. The rest found that the amount of radon reaching the living areas of their homes wasn't nearly as high as they had feared. They gained some valuable information about their houses (at least one couple changed plans to convert part of their basement into a playroom for their children, basing their decision on the test readings for that part of the house) and discovered that there was no reason to live in fear of radon.

7

GETTING RID OF RADON

The odds are overwhelmingly in favor of your house receiving a clean bill of health from a thorough radon testing program. But what if you aren't so lucky and yours is the one house in ten with a real problem? When faced with such a basically alien situation—you may, after all, have never even heard the word "radon" until a few months ago—it's not unusual to feel frightened, confused, and even angry. But if you find yourself there, just remember this: There hasn't been a house found yet with a radon problem that couldn't be fixed.

Most radon contaminations are relatively minor; they're at, or not far over, the Environmental Protection Agency's (EPA) 4 picocuries per liter (pCi/l) action level. While not to be ignored, these problems can often be fixed with a minimum of disruption and expense. In some cases, you may be able to do some or all of the work yourself and keep costs down even more.

Serious problems may require more time, trouble, and expense to remedy. But they can be remedied, as

WHAT TO DO IF THE NEWS IS BAD

In considering whether and how quickly to take action based on your test results, you may find these EPA radon guidelines useful. They're based on the agency's recommendation that you should try to permanently reduce your levels to about 0.02 working level (WL) or 4 picocuries per liter (pCi/l)—or below—in the living areas of your house. The higher the radon level, the faster you should initiate action.

If your results are about 1 WL or higher, or about 200 pCi/l or higher: Exposures in this range are among the highest observed in homes. Residents should undertake action to reduce radon as much as possible. It is recommended that action be taken within several weeks. If that can't be done, consider temporarily relocating until the levels can be reduced.

If your results are about 0.2 WL to about 1 WL, or about 20 pCi/l to about 200 pCi/l: Exposures in this range are considered seriously elevated. Undertake remedial action to reduce levels within several months.

If your results are about 0.02 WL to about 0.1 WL, or about 4 pCi/l to about 20 pCi/l: Exposures in this range are considerably above average. Residents should undertake action to lower levels to about 0.02 WL (4 pCi/l) or below. It is recommended that you take action within a few years, sooner if levels are at the upper end of this range.

If your results are about 0.02 WL or lower, or about 4 pCi/l or lower: Exposures in this range present some risk of developing lung cancer. However, reductions of levels this low may be difficult, and sometimes impossible, to achieve.

experience with a wide variety of radon problems in practically every type of house has shown. And although a house with very high concentrations of radon could cost thousands of dollars to fix, it's not always necessary to do all the mitigation work at once. Some homeowners choose a multiphase repair program, working in stages to steadily reduce radon more and more. While not ideal, a gradual reduction process is better than none at all.

RADON INSPECTION TOUR

The first step toward fixing a radon problem is finding out where the radon is coming from. The more you know about how it's getting into the house, the more you can do to either keep it out, or get rid of it once it's indoors. Until you have at least some of the answers in hand, trying to cure the problem can be a real hit or miss proposition.

You probably already have a general idea of your radon sources, based on what you've learned so far about how radon behaves and from the results of the follow-up testing you did in various parts of the house. But it's often helpful to take an even closer look at your house, to analyze it objectively and inspect it perhaps more closely than you ever have before.

You'll be looking for the pathways that this invisible, odorless gas follows into your home. Notice I said *pathways*. If there's radon in the soil under and around your house, it's likely that the gas has found more than one way in. So don't stop when you find one potential entry, such as a badly cracked wall or an open drain. Radon gas travels the path of least resistance. If you plug one entry it may simply mean that more radon will come in through another point. It's not unusual, say radon mitigation contractors, to identify and close an opening

SOURCES OF RADON IN HOMES

Source	Estimated Contribution
Building materials	0.3 to 30 pCi/second
Soil gas diffusion	3 to 6 pCi/second
Soil gas transport (cracks, block walls)	0 to 150 pCi/second
Tap water	60 pCi/second for water with 10,000 pCi/l of radon

Source: United States Environmental Protection Agency, Office of Radiation Programs.

responsible for 50 percent of the radon coming into a house and then find little or no overall reduction because the radon has found another way into the house.

START IN THE BASEMENT

So, with clipboard and pencil in hand, conduct your inspection. The logical place to start is the basement.

The Pennsylvania Department of Environmental Protection's Bureau of Radiation Protection has compiled the following helpful list of questions to answer as you make your basement tour:

1. Does the basement have a poured concrete floor throughout, or are there sump holes, a root cellar, an unpaved floor, or any exposed earth?
2. Does the basement floor have any penetrations (downward-flushing toilet, heating system water pipes, sewer lines, or floor drains)?

3. What is the condition of the basement floor? Are there visible cracks? Have concrete joints separated?

4. What kind of perimeter, load-bearing walls does your basement have? Are they cinder block? Poured concrete? Stone?

5. Does any basement perimeter wall have below-grade penetrations (water or sewer lines, pipes in sleeves, breaks, holes, or cracks)?

6. Do basement perimeter walls show signs of water penetration (an irregular, powdery white marking or moisture near the bottom of the wall)?

Also, be on the lookout for heating ducts below a slab, any obvious pathways from the soil into the basement, an unvented crawl space, or the presence of drafts that may indicate that outside air is entering.

Your inspection may leave you with a clear idea of how radon is getting in. It may also tell you that you need to run a few more short-term charcoal tests to nail down your suspicions about possible entry points. The easiest way to test a specific area is to place a charcoal tester nearby and build a tent around it by taping a sheet of plastic to the wall or floor. Or, if the area is small, you can upend a wastebasket over it and the tester and seal the edge with temporary caulking (the press-on type used to weather-strip windows should do the trick). This additional testing will probably only verify your original hunch, but if it makes you feel more comfortable to have "proof," go ahead and test.

RADON HOUSE DOCTORS

It's possible that your house inspection will only leave you more confused about the source of your radon. If so, consider calling in professional help. The need for expert consultation has spawned a new breed of house investigators, known as *radon house doctors.* These consultants use many of the same high-tech tools that specialists in energy improvement use to help homeowners locate heat leaks and air-infiltration points. But while many radon house doctors may have gained their basic skills evaluating the energy efficiency of houses, they must also be knowledgeable about how radon behaves in order to diagnose the source of problems.

For a fee that usually starts at between $50 and $100, a radon house doctor will visit your house, probably start the investigation with a visual inspection, and may take grab sample measurements at suspected source locations or suggest additional places for you to test yourself with charcoal detectors.

Most radon house doctors will also bring in some special equipment to measure infiltration rates to help determine their influence on your indoor radon levels. Among the most commonly used instruments are smoke tracers, pressure gauges, air velocity indicators, and blower doors.

SMOKE TRACERS

Watching how strategically placed puffs of smoke behave can tell a lot about the flow of air in a house. For example, if a bit of smoke released near a floor or wall crack flows quickly away from the opening, there's probably air coming in through it. Smoke flowing into the crack means a pressure differential is pulling air out of the house through the opening. Timing the movement of the smoke over a fixed distance will also provide information on the strength of the infiltration.

Many house auditors use a chemical "smoke" from titanium tetrachloride or sulfur dioxide, squeezed from a tube with a rubber bulb. Lighted cigarettes, matches, or other combustion sources of smoke could pose a fire hazard and are not recommended.

PRESSURE GAUGES

Various types of pressure gauge, some very complicated, others relatively simple, can be used to measure differences in air pressure in different parts of the house. As we've seen, when the air pressure in a basement is lower than that of the outside, the difference can cause more radon gas to be sucked indoors.

AIR VELOCITY INDICATORS

Devices that measure the speed and volume of airflow could prove useful to verify movement of drafts and infiltration into the house.

BLOWER DOORS

One of the most common tools of energy doctors, the blower door, is a large fan assembly that temporarily replaces the front or back door of the house. When the fan is started, it sucks air out of the interior and creates a higher negative pressure in the house, which exaggerates infiltration leaks. The technician may then walk around with a smoke stick to see where the air is coming in, or inject the house with Freon gas and use Freon-detecting equipment to follow its movement through the house.

Again, these are the very same tools used by home energy auditors, but radon source identification is a very different game, requiring special experience and knowledge. For help in finding the right person or company for the job, contact your regional EPA office or the agency that handles radon questions in your state. Be wary of any company that uses high-pressure sales tactics or wants you to sign on the dotted line after the first contact.

And while most radon house doctors are undoubtedly honest and well meaning, it's probably safest not to hire the same people who diagnose your problem to fix it. You want the testing process to be independent, not based on future profits for a contractor. Ask your radon doctor for advice on remedial action, but get bids from several companies for the work you finally decide to have done.

PHASING OUT RADON

"Unfortunately, there's no easy cure for a lot of radon problems," says Gene Durman, senior policy advisor for the EPA. "There aren't any radon-destroying machines that simply make it go away. At this point we can't say to homeowners, 'Do this and the radon will go away.' Every house is different, so a number of approaches were developed."

The EPA, state public health agencies, and even some private companies, have been looking intensively into radon reduction techniques for several years now. Standard procedure has been to find a house (or houses) with radon problems and use it as a laboratory to try out reduction methods. For the trouble of living with workmen and radon monitors, the homeowners get a free or low-cost fix.

It's not only badly contaminated houses that are studied, either. Usually, buildings with a range of radon levels are chosen, since different degrees of radon require different repair methods.

The legacy of the testing done so far is an impressive body of data on how radon behaves and how it can be controlled. While no one is offering any guarantees that the methods used will work in every house the way they did in the test houses, costs, techniques, and average reductions have been fairly well documented.

The bottom line is that, depending on your situation, you can deal with radon in one of two ways: by preventing it from entering your house in the first place or by removing or diluting radon once it has gotten indoors. There's more than one way to keep radon out of houses, and sometimes it takes a combination of more than one strategy to do the job completely. The methods include the following:

- Sealing cracks, crevices, loose joints, penetrations, exposed earth, or any opening that allows the flow of radon into the house
- Diverting the source of radon from the interior of the house by installing a ventilation system that can actually pull radon-laden air out from under the slab, from inside the cinder block walls of a basement, or from the soil around the perimeter of the foundation, and exhaust it harmlessly into the outdoor air
- Reversing the pressure differences between the inside and outside of the house so that air flows from indoors to outdoors
- Changing a water supply that is bringing radon into the house or filtering the water to remove radon
- Replacing or sealing building materials that are giving off radon gas

If it's not practical, or possible, to cut radon off at the source, you've got to deal with it once it's gotten indoors. And the only effective way to do that is by replacing or displacing radon-contaminated air in the basement, crawl space, or living area with an equal amount of fresh air from either natural or forced ventilation.

Since radon reduction isn't always an exact science, protecting yourself will usually require a combination of the two approaches to reach a point you can feel comfortable with. Or you might want to tackle the problem by starting with some of the low-cost remedies that you can accomplish yourself, then testing the house again to see how far they've taken you. That may be as much as you need to do. If the levels are still unhealthy, you can go on to the next phase and bring them down further.

Here's some advice about attempting the more complicated remedial projects on your own: Think long and hard about it first. Radon removal can be a complicated business and often requires special construction skills. At the very least, hire a contractor to consult with you on the job. If you're an experienced do-it-yourselfer, or work in the building trades, you may be able to do some of the work yourself and save a considerable sum on labor costs.

"We've looked at some owner-installed systems that were very well done," says Gene Tucker, of the EPA's Air and Energy Engineering Research Laboratory, and chief of the Indoor Air Branch. "We've also seen a couple of systems that didn't work because some of the work was done sloppily."

"I personally think it's great when people try to do some of the work themselves, and I hope we're getting information to the handy homeowner who wants to try sealing cracks or installing fans," he continues. "But, if possible, they should have a professional or a state agent check out their work afterwards."

However, before you go jackhammering holes in your slab, or installing venting pipes in basement walls, make sure you've got a detailed blueprint in front of you, and some guidelines from a contractor experienced in

radon reduction behind you. As said, you may be able to
hire a professional on a consulting basis to design the
system and oversee your work.

The rest of this chapter will outline the methods that
have proven effective at reducing radon levels in test

HOW TO HIRE A CONTRACTOR

Before you sign on with a contractor to fix your radon
problem, you should do some research into the company's background and past performance. Because in the
long run, that's about all you're going to have to go on.
In the world of radon remediation, there's almost no
such thing as a guarantee. You can insist that the contractor warrant materials and workmanship, but you
are unlikely to get a guarantee that when the work is
done, radon levels in the house will be below a speci-
fied number. Radon problems are so unpredictable,
and houses so variable, that few contractors are willing to stake their fees on being able to fix them.

Here are some tips on how to find a contractor
who knows what he or she is doing and who has a
proven track record with radon-related remedial work,
and how to avoid unqualified or unscrupulous firms:

- Find out how long the company has been in business in your area and where the office is located.
- Ask the company to provide the names of three
  customers it has done similar work for, and then
  you should give *each one* a call. Ask the clients
  about their overall satisfaction with the contractor and about any problems that arose in the
  remedial work. Find out by how much radon
  levels were reduced and what methods were em-
  ployed.

houses around the country. It's based mainly on experiments conducted by the EPA's Office of Research and Development and Office of Radiation Programs, the Pennsylvania Department of Environmental Resources, the New York State Energy Research and Development

- Ask for municipal building inspector references.
- Check with the Better Business Bureau or Chamber of Commerce for records of any complaints.
- Check with state or local trade organizations that the contractor may belong to.
- Contact the agency in your state responsible for radon information (see Appendix A) and ask if the contractor has completed EPA or state educational programs for radon mitigation.
- Get a second opinion. Reports have surfaced in some parts of the country of contractors charging highly inflated prices for radon remedial work. When the homeowners question the cost, the contractors sometimes cite a General Accounting Office (GAO) "report" that quotes amounts ($4,300 to $15,700) that the GAO spent fixing some experimental houses in Pennsylvania. The EPA says the GAO's numbers are far higher than what the average person should have to spend. In any case, estimates should never come out of reports and guidelines; they should be based on the situation at hand. Obtaining several written estimates from different contractors, each spelling out the work that will be done, will help you make the right choice and avoid being victimized.

Authority, the Pennsylvania Power and Light Company, and Bill Brodhead, a Bethlehem, Pennsylvania, home builder and radon mitigation contractor.

Along with a detailed description of each method, you'll find the approximate cost to buy and install the

A SHORT-TERM FIX

There are ways to reduce your risk from radon immediately. While not permanent fixes by any means, these actions can help protect you from additional exposure to high levels of radon until you decide on a long-term solution.

- Stop smoking and don't let anyone else smoke inside your house. The tiny particles in cigarette smoke can make indoor radon more dangerous.
- Don't spend a lot of time in the basement, or anywhere in the house where radon concentrations are high. Close off those areas from the rest of the house with doors or partitions.
- When the weather permits, ventilate the house as much as possible. Leave windows or doors on both sides of the house open to allow cross-ventilation, and use fans that blow air *into* and not out of the house.
- If your house has a vented crawl space, open all the vents on every side of the house to keep air moving. This should help to draw off some radon gas before it gets indoors.
- Avoid depressurization of the house by providing furnaces and wood stoves with outside air supplies. Don't use the fireplace, and keep the chimney damper closed. Keep attic fans or large kitchen exhaust fans turned off.

REMEDIAL READING

Before you start remedial action to lower radon levels in your house, send for the Environmental Protection Agency's (EPA) *Radon Reduction Techniques for Detached Houses: Technical Guidance.* This 50-page booklet, filled with illustrations and construction details, should be required reading for anyone planning to undertake any of the radon mitigation techniques described in this chapter. The guide is available free by writing to the following: Center for Environmental Research Information, Distribution, 26 West St. Clair Street, Cincinnati, OH 45268.

equipment (based on professional installation), and, where it applies, the yearly cost of operating the system. Also listed is the estimated annual average reduction in radon concentrations of each method, based on the measurements taken in the test homes it was used on.

HOW TO USE VENTILATION TO REDUCE RADON

Given radon's unpredictable nature, and the many ways it finds of getting into some houses, it's often easier to keep as much radon out of the house as you can, and then deal with the rest through ventilation. As you'll see, proper ventilation techniques can reduce radon levels by a great deal—90 percent or more in some cases. But as you'll also see, ventilation, especially without some sort of heat recovery, can pose comfort problems that some people may find hard to live with.

NATURAL VENTILATION

Estimated average reduction: Up to 90 percent
Cost of installation: None
Yearly operating cost: Your heating and cooling bills
    could triple

There's nothing very complicated about natural ventilation. It's about opening some extra windows. When you do that, air containing radon will either be continually swept out of the house or greatly diluted by incoming air.

The average American house, which can hardly be described as tight, goes through at least one full air change per hour (ACH), meaning that all the home's indoor air leaks away at least once every 60 minutes, even with all the doors and windows closed. Some drafty older houses may go through 2 ACH, and some new, tightly constructed ones typically undergo 0.5 ACH, although some *very* tight ones have gotten down to 0.1 ACH, or one-tenth of an air change per hour.

If you have a basement, it can be naturally ventilated by opening windows on all sides of the house to allow a flow-through of air. If your house has a crawl space, open *all* the vents. If the house sits on a slab, your only choice is to open windows and doors in the living space.

Houses that already have high air-exchange rates probably won't be helped a lot by extra ventilation. But tighter houses can benefit. For example, doubling the ventilation from 0.25 to 0.5 ACH can halve the amount of indoor radon. Increasing the ventilation rate still further, from 0.25 ACH to 2 ACH can reduce radon levels by as much as 90 percent. However, there are limitations to this approach. It works well in houses with lower amounts of radon but by itself won't be enough to bring very serious contaminations (above 40 pCi/l) under control.

And obviously, there are some comfort and economic limitations. Opening windows may be no great sacrifice when the weather is mild, but it's hardly practical during the winter in most parts of the country. In cold climates, a 60°F or cooler basement is bound to affect temperatures in the rest of the house. Shutting off the basement and using it as little as possible will help, but your heating bills will inevitably be higher. How much higher? It depends on where you live, but you could see your costs double or triple during the deep winter months (based on a 40 percent increase in the basement ventilation rate). The same could hold true for cooling bills during the hot parts of the year.

In addition to the costs, there are comfort issues— cold floors, drafts, and a frigid basement you probably won't want to spend much time in. And electric "heat tapes" may be necessary to keep pipes from freezing.

Realistically, natural ventilation isn't the final answer for most houses. It can't be used in most parts of the country during the winter, and the cold months are when radon levels coming from the soil are often at their highest. So, at best, it's a short-term fix for houses with low to moderate radon contaminations.

ADVANTAGES
OF NATURAL VENTILATION

- A quick fix for less serious radon contaminations
- Costs nothing to install

DISADVANTAGES
OF NATURAL VENTILATION

- Can significantly increase heating and cooling bills
- May cause necessity of adding insulation to thermally isolate basement from rest of house
- Good for only part of the year in slab-on-grade houses

FORCED VENTILATION

Estimated average reduction: Up to 90 percent
Cost of installation: $150 for fans
Yearly operating cost: $100 to operate fans, plus an up to
    threefold increase in heating or cooling bills

Forced ventilation with window fans works the same way as natural ventilation, but insures a steadier and more reliable ventilation rate. It may, for instance, be hard to get the ventilation rate in a tight house to a target ventilation rate of 2 ACH without using fans.

The ventilation and distribution of air can be controlled by the number and location of the fans and by using fans that have louvers that allow you to direct the airflow. You'll probably need a minimum of two heavy-duty window fans, each capable of moving air at a rate of at least 240 cubic feet per minute (CFM). Place the fans in basement windows and upstairs in the living space when the weather permits, and open other windows for cross-ventilation.

Make absolutely sure that the fans are installed so that they're bringing in fresh air from outdoors and not exhausting indoor air, which could create a negative pressure inside and draw more radon from the soil. Even properly installed basement fans can cause some depressurization if the air entry and exhaust points aren't well balanced. But what's well balanced in one house might not be in another, so there's no way to tell you exactly where the fans should go in your house and which corresponding windows should be opened. The only way to verify the effectiveness of your setup is to retest the house with the fans in operation.

The fans shouldn't cost you more than $150 to buy, and if you're installing them in a crawl space you may need to pay an electrician to run some new wiring. The fans will probably cost about $100 in electricity yearly to operate, and you can expect the same increase in heating and cooling bills as with natural ventilation, perhaps even higher. For example, increasing your air exchange to eight times its normal rate, more than possible with forced ventilation, will more than triple your present heating or cooling bill.

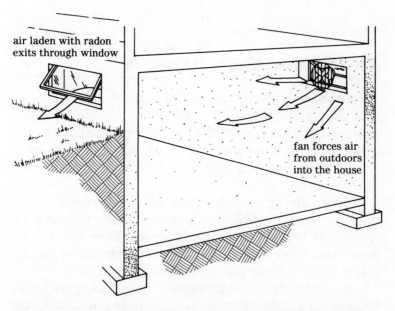

air laden with radon exits through window

fan forces air from outdoors into the house

Forced air ventilation with fans can be used in a basement or living space to remove or dilute radon gas to safer levels. However, it's important to open other windows to create flow-through air movement. This approach can be effective in many cases, but it's not very practical during cold weather.

ADVANTAGES
OF FORCED VENTILATION

- The same as those of using natural ventilation, but reductions may be greater

DISADVANTAGES
OF FORCED VENTILATION

- Fans may be noisy
- Will increase both your electric and heating and cooling bills
- Basement may no longer be usable as living space

AIR-TO-AIR HEAT EXCHANGERS

Estimated average reduction: 96 percent
Cost of installation: $400 to $2,000
Yearly operating cost: $75 to $100 for electricity to run the fans, plus an increase of at least 30 percent in heating and cooling costs

Air-to-air heat exchangers offer the radon reductions of forced ventilation without such enormous increases in heating and cooling costs. They do that by recovering more than half of the heat from indoor air and transferring it to incoming fresh air.

Available in room-size models, which look much like window air conditioners, or whole-house units, air-to-air heat exchangers (also known as heat recovery ventilators) have been in use in energy-efficient houses for many years. Although designs vary from maker to maker, all work on the same basic principle: One or more blowers inside the unit move two separate airstreams past each other inside a heat exchanger—an exhaust stream of stale indoor air and an intake stream of fresh but cold outdoor air. Heat passes from one stream to the other, cooling the stale air and warming the fresh incoming air. The heat exchanger itself may be made of plastic, paper, or metal, but it should be capable of passing the heat between the two airstreams without them mixing. Otherwise, the fresh air coming in will be contaminated by the radon in the outgoing air.

Unless radon concentrations in a home are very low, a room-size air-to-air heat exchanger probably won't be enough. Its air-moving capacity is too low to ventilate large areas, so putting one at one end of your house may not do much for the air quality at the other end, especially if there are many doors, halls, and other impediments to airflow.

Whole-house models are designed to be mounted in a home's basement or crawl space and run automatically. In addition to the machine itself, you'll also have to add some ductwork to carry the air to and from the heat exchanger, which could add $500 or more to the cost. At least four ducts will be necessary to distribute the system's air. One carries fresh air from outdoors to the unit, another carries fresh air from the unit to the house, a third carries stale air from the house to the air exchanger, and the fourth exhausts stale air outdoors.

Existing basement windows can usually be used for collecting and exhaust vents. But when planning your setup, keep in mind that if the discharge airstream is too close to the fresh air intake, the system will "short circuit" and you'll be drawing polluted air back in with the fresh. The two ducts should be placed *as far away from each other as possible*, preferably on opposite sides of the house.

An unbalanced airflow, caused by duct runs that are too long or ducts that are too small, can also affect air-to-air heat exchanger performance. When you buy a unit, make sure it comes with clear instructions that show you how to size ducts and calculate maximum duct runs. Or, buy the system from a distributor that will install it.

Air-to-air heat exchangers have proven to be very effective at reducing radon levels in tightly built houses where the original concentrations were below 40 pCi/l. Researchers at the Lawrence Berkeley Laboratory showed that in an energy-efficient test house in Carroll County, Maryland, radon levels fell in direct relation to increased ventilation rate. In another test, heat recovery ventilators in a basement were able to reduce radon levels by 96 percent. Based on these and other tests, the EPA says that "confidence in the effectiveness of this technique should be high."

The trick to using the technique, though, is in getting an air-to-air heat exchanger to provide enough ventilation to significantly reduce radon levels in your house. That means sizing the unit you buy to the square footage of the area to be ventilated. If, for instance, you want to ventilate a 30-by-30-foot basement with 8-foot ceilings, the total volume is 7,200 cubic feet. If you want 2 ACH, you need to ventilate 7,200 × 2, or 14,400 CFM. Since ventilators are rated in cubic feet per minute, you divide the cubic feet per hour by 60 to figure the size fan you'll need, in this case, 240 CFM. If your basement's original air-exchange rate was 0.25 ACH, which is possible in a tightly constructed house, raising it to 2 ACH could reduce radon levels by 90 percent.

But the question seems to be, can you trust the cubic-feet-per minute ratings the manufacturers put on their air-to-air heat exchangers? When *Rodale's New Shelter* magazine (now called *Rodale's Practical Homeowner*) tested six brands of air-to-air heat exchangers in 1984, they found that all but one averaged 39 percent lower airflows than their specifications promised. Based on the test results, the magazine cautioned readers that "it's wise to assume that an air-to-air heat exchanger will actually deliver only about 0.7 of its rated airflow."

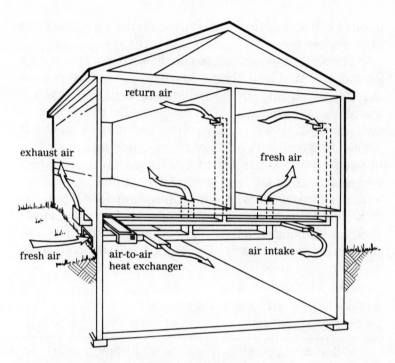

An air-to-air heat exchanger installed in a basement or crawl space can provide fresh air to the house while recapturing 70 percent or more of the heat from the outgoing air. Installation costs can run up to $2,000.

New Shelter also found that the units it tested were, on average, about 70 percent efficient when run on their highest fan speeds. In other words, they captured all but about 30 percent of the heat from outgoing house air. That's pretty impressive, and certainly far more attractive than using fans with no heat recovery to ventilate your house. But since you'll probably have to keep your air-to-air heat exchanger running 24 hours a day, your heating and cooling bills can be expected to rise about 30 percent after you install one. Additionally, figure on about $100 a year in electricity to run the fans inside the heat exchanger.

There is a potential problem with air-to-air heat exchangers that you should know about before you decide to install one: In cold climates, they can chill the warm, moist exhaust air from your house below its condensation point; water may collect inside the unit and in very cold weather can freeze there, blocking the airflow. In one survey, 40 percent of 147 Canadian homeowners who used air-to-air heat exchangers said they had experienced problems with winter freeze-ups. So if you live in a cold area of the country, consider buying a model with an automatic defrost feature, or one that has special drains that allow you to siphon off collected moisture.

ADVANTAGES OF AIR-TO-AIR HEAT EXCHANGERS

- Recover a large percentage of a home's heat while providing continuous fresh air
- Dehumidify house

DISADVANTAGES OF AIR-TO-AIR HEAT EXCHANGERS

- Can be somewhat noisy (but no more so than a window fan)
- Will increase heating and cooling bills by around 30 percent
- Even with heat recovery, the airflow into living space will create drafts that may make people feel uncomfortably cool
- Whole-house models are difficult to install in finished basements because of the extra ductwork required

AVOIDING HOUSE DEPRESSURIZATION

Estimated average reduction: 0 to 50 percent
Cost of installation: Varies
Yearly operating cost: Minimal, could even *save* you
    money

Some household appliances, among them clothes dryers, furnaces, gas water heaters, fireplaces, and wood stoves, consume indoor air while they're operating and exhaust it outdoors. When all these appliances are operated at the same time, they can exhaust as much as 600 cubic feet of indoor air per minute.

When substantial amounts of indoor air are drawn into an appliance, that contributes to creating a slight vacuum inside the building. And as we've seen, this lowering of the air pressure, called *depressurization,* can cause more radon-laden air to be drawn into the house through cracks and openings in the basement or slab.

You don't have to stop using your furnace or pull out your wood stove and dryer to avoid depressurization. But you do have to see that they get another source of air, one from outside the house.

Actually, the American Society of Heating, Refrigerating and Air-Conditioning Engineers (ASHRAE) has been touting outside combustion air for furnaces and wood stoves for more than five years. They say that when all these air-consuming appliances are competing for indoor air, there may not be healthy amounts of fresh air left for the people in the house. And experts have long recommended outside air supplies for wood stoves and fireplaces, because the extra air helps them burn their fuel more completely.

Avoiding depressurization by supplying outside air sources to appliances won't cure most radon problems, but it may help. Past testing has indicated that reductions range from 0 to 50 percent when makeup air is provided to furnaces and fireplaces, and 0 to 10 percent when outside air is supplied to clothes dryers, exhaust fans, and any appliances used no more than 20 percent of the time over the course of a year.

But, as you'll see under "Block Wall Ventilation," supplying outside air can also help increase the effectiveness of other reduction methods, specifically cinder block wall suction systems.

To provide makeup air to an appliance, you'll need to run an insulated fresh air duct, outfitted with dampers to keep cold air from coming in when it's not needed, from an exterior wall of the house to the air intake of the appliance. In many cases, installing fresh air inlets requires a custom approach—you'll have to *create* intakes where none existed and block the appliance's normal vent or inlet.

Installation will also mean breaking through basement or exterior house walls, and, with fireplaces, drilling holes through brickwork. It all sounds like a job for a professional heating contractor, who'll also be knowledgeable about sizing fresh air ducts and general safety. At least call someone in to help you design your system, or cut costs by hiring a pro to do the rough work and do the finish work yourself. For additional guidance on installing outside air supplies for combustion appliances, send for the report, *Supplemental Combustion Air to Gas-Fired House Appliances* (DOE/CE/15095), available from the United States Department of Energy, Washington, DC 20585.

ADVANTAGES OF AVOIDING HOUSE DEPRESSURIZATION

- Usually reasonably inexpensive
- Little or no operating costs
- Can be installed using inexpensive, readily available materials

DISADVANTAGES OF AVOIDING HOUSE DEPRESSURIZATION

- Effectiveness difficult to predict
- Reductions are generally small

A WORD ABOUT AIR CLEANERS

We've all seen the advertisements for air filters and electrostatic precipitators (sometimes called negative ion generators), machines that remove particles and dust from indoor air. Inexpensive, tabletop air cleaners were the rage a few years ago—you could hardly turn on the television without seeing an ad for still another "solution to indoor pollution."

Recently, with all the publicity surrounding radon, we've been hearing about them again. The hope was that these air cleaners could be used to reduce the effects of indoor radon.

Unfortunately, that doesn't seem to be the case. "Existing information does not clearly document the effectiveness of air cleaners in reducing the risk of developing lung cancer," says the EPA. "Until more is known, EPA believes that the data do not warrant discontinuing the use of air cleaners already installed, nor can we suggest installing air cleaners to reduce the risks."

Behind that official-sounding statement is some controversy. Indeed, tests have shown that electrostatic precipitators and air filters *can* reduce radon working levels, sometimes by 50 percent or more. In both instances, the air cleaners worked by removing particles from the air, some of which had radon daughters attached.

The trouble is, when there's radon present in a house, only some of its daughters become attached to dust and smoke particles. The rest remain unattached, and they can be even more dangerous when inhaled. Air cleaners can do nothing to remove these unattached daughters from the air; they reduce working levels by removing particles, but may do very little to lower your overall risk of lung cancer from radon.

So, until scientists know more about the relative risks of attached and unattached radon particles, using air cleaners to treat indoor radon problems may not be advisable.

HOW TO KEEP RADON OUT OF THE HOUSE

In some houses, added ventilation may be the best and cheapest solution to a radon problem. In tight houses with low to moderate levels, that's all it takes. But, for others, the answer lies beyond simply bringing more air indoors. "We've come across many homes where the amount of radon entering the house would take an impractical amount of ventilation," says Harvey Sachs, an indoor pollution expert with the National Indoor Environmental Institute. "The best solution in these cases is to track down the main entry points," he points out.

Track them down and seal them—or eliminate them—or divert radon gas away from them. The following radon cures are all ways to make sure that radon never gets into your house. These are the heavy-duty solutions, the ones that have been used to fix the most seriously contaminated houses. Some of the methods are expensive; some are complicated. But together, and sometimes by themselves, they work.

SEALING ENTRY POINTS

Estimated average reduction: 30 to 90 percent
Installation cost: $300 to $500
Yearly operating cost: None

As simple as it may sound at first, sealing your basement or slab off from the radon in the soil under and around it is a formidable task. As the University of Pittsburgh's Dr. Bernard Cohen puts it, "Picture your basement surrounded by water. You've got to fix it so that no water could get inside. And that's not easy, because it seems like there's always one more crack the water could get in through."

The best thing about this approach is that you can do it yourself. The worst thing is that you may find yourself doing it for a long time.

The best place to start is by sealing the visible cracks you can find in the floor and walls. The instructions for tightly sealing those cracks were developed by the Penn-

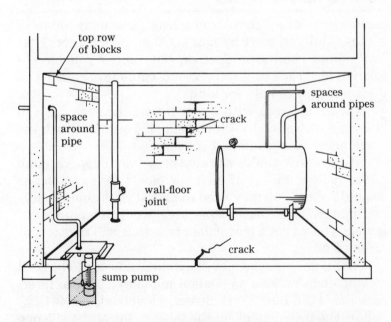

These are some of the radon entry points in a basement that will have to be sealed if you want to reduce radon levels. Sealing cracks is time-consuming work, but it's not expensive and is sometimes effective at reducing radon to safer levels.

sylvania Department of Environmental Resources (DER). They call for the following tools:

- Cape chisel (a chisel with a tapered blade, used to cut grooves and for work in tight spots)
- Caulking gun
- Fan to ventilate the basement while you work
- Masonry hammer
- Masonry trowel
- 1-inch cold chisel
- Safety goggles
- Wire brush

HOW TO SEAL CRACKS

Sealing cracks so they won't leak is usually done in stages. And you start by making the crack bigger. That may sound counterproductive, but unless you have a fairly wide crack to fill, the grout, caulk, or sealant you fill it with won't hold for long.

Clean the crack with a wire brush, then enlarge it to at least 1 inch wide by 1 inch deep, if you can. Brush the crack again or vacuum out the dust.

If the crack you're working on is very deep, through to the hollow middle of a cinder block, for example, or around a pipe, you may need to first stuff in some "backing" materials like newspaper, steel wool, or oakum. Otherwise, you've got a bottomless hole that will take lots of caulk to fill.

To fill floor cracks, use "flowable" silicone (The DER recommends Vulkem 45 Sealant made by Mameco International, 4475 East 175th Street, Cleveland, OH 44128.) Follow the instructions on the label to inject the silicone into the crack to a depth of ¼ inch. Let it cure for four days; after that you shouldn't be able to lift it out. If you can, the surface probably wasn't clean or wide enough when you started.

Wall cracks are prepared the same way, but should be filled with a nonflowable silicone caulk, such as those made by General Electric and United Gilsonite Laboratories (available at most home centers).

When filling spaces around a pipe, post, or utility penetration, chisel out a 1-by-1-inch channel around it. Then brush any exposed pipe and clean it with paint thinner. When the thinner is dry, stuff in backing material as needed and fill the crack with the appropriate silicone, depending on whether the pipe is through the wall or floor.

After all your silicone fillers have cured, covering them with an epoxy sealant and a nonshrink grout and bonding mix will help make a tighter barrier against radon gas. Brush on the epoxy (the DER suggests Sikadur 32 Hi Mod, from Sika Corporation, P.O. Box 297, Lyndhurst, NJ 07071) as directed on the label, to the sides and top of the silicone and the pipes. Then, when the adhesive is still tacky, trowel the grout mix (the DER recommends Sikatop 122, also by Sika Corporation) flat over each filled crack. In some EPA tests, asphalt roofing cement proved an effective sealer for spaces around pipes.

Radon can sometimes flow right through a porous concrete floor or a basement wall made of cinder blocks. Applying a barrier coating of a two-component epoxy resin paint to the walls will help to seal the blocks. You'll find the epoxy paint at most major paint distributors. It costs about $30 per gallon, and you'll probably need at least two coats. A good waterproof latex paint can slow down or stop the gas if the walls or blocks are concrete and not particularly porous. After filling cracks using the methods above, thoroughly clean and brush the floor or wall and apply two or more coats of the sealant or paint. The results of one study of this method, published in *Health Physics*, showed it reduced radon concentrations by 87 to 97 percent on the inside of the wall.

Masonry block walls also have the bad habit of acting as "chimneys" that funnel radon gas into the house from the soil below. Stacked one on top of each other, their hollow centers provide a perfect conduit. To make matters worse, some builders neglect to cover or plug the holes in the top run of blocks on basement walls of the houses they build. So the radon is free to travel up the voids in the blocks and directly into the basement or up into the living space.

Needless to say, if that's the situation in your basement, you'll want to take steps to change it. If the top row of blocks is completely uncovered, and you can access them, stuff crumpled newspapers down into each void. Then fill the entire hole with a 2-inch-deep layer of mortar.

If there's not enough room to maneuver a trowel, stuff in the newspaper and fill the void with a single-component urethane insulation foam (available in spray cans at hardware stores). But if you have a lot of blocks to fill, you may want to look into bulk kits like those sold by Fomo Products, 2775 Barber Road, Norton, OH 44203.

In some houses, the void may be uncapped, but entirely covered by the sill plate. If *no part* of the void is visible under the sill, you can fill the seam between the sill and blocks with silicone caulk.

HOW TO SEAL SUMP HOLES

Sump holes in basement floors present some thorny problems. If the sump is rarely needed, consider getting rid of it and sealing the hole. To do that, fill it with clean gravel to within 6 inches of the basement floor. Then level and tamp the gravel. Next, place a sheet of 6-mil polyethylene over the gravel as a vapor barrier. Add 2 inches of sand over the plastic and clean the edges of the existing concrete. Apply an even coat of epoxy sealant to the edges, and, without delay, pour a layer of concrete over the sand (you must cover the hole in one pour so that there are no joints). Screed the concrete level, then finish it smooth with a metal trowel and feather the edges. After the concrete has cured, apply flowable silicone along the joint between the old and new cement. (This same process can be used to cover and seal any exposed patches of earth in your basement.)

If you absolutely must keep the sump pump, the next best solution is to isolate it from the rest of the basement and vent back to the outdoors the radon gas that comes up through the hole (see the illustration on page 169). You can make the sump cap out of sheet metal as long as all joints and edges are carefully sealed. A small electric fan should be installed on the end of the polyvinyl chloride (PVC) exhaust pipe to suck out radon gas. You should be able to build this basic sump ventilation system yourself for about $100.

Make the same kind of tough evaluations of any floor drains in the basement. Do they ever get used? If not, cement them up. Or better, see if you can get tight-fitting covers that will allow you to use the drains in an emergency.

HOW TO SEAL A CRAWL SPACE

New houses built over crawl spaces are quite rare these days, but there are still a lot of older versions of these houses around. If yours is one, you know that none of the above applies to you. You probably won't get too excited over the idea of paving the exposed earth under your house—that's all there is down there. But what you *can* do is try to prevent the radon gas from coming up from the earth and through your floor into the house.

If you don't have vents all around the perimeter of the crawl space, install them and think about adding some kind of forced ventilation system. Installing an air vapor barrier of polyethylene or PVC under the floors will cut infiltration of the gas, as will sealing any openings between the floor and the crawl space. Results of house tests conducted by researchers at Simon Fraser University in Canada suggest that the combination of a ventilated crawl space and "bottomside" vapor barriers can result in radon reductions of about 60 percent.

No matter what you do to try and seal radon out of your house, be sure to run another follow-up test to check the results. And even if that test shows you've brought amounts down to an acceptable level, don't forget about radon forever. Even the best caulks can shrink and dry out with age; foundations may continue to shift and settle, opening new cracks. Retest at least every year as insurance.

ADVANTAGES OF SEALING ENTRY POINTS

- Can be a do-it-yourself job
- Relatively inexpensive
- Usually no operating costs
- Increases performance of additional mitigation methods
- Materials are easy to find

DISADVANTAGES OF SEALING ENTRY POINTS

- Often impractical for finished basements
- Cracks not always visible
- Not permanent
- Reductions are sometimes small
- Crawl spaces make very cramped workspaces

DRAIN-TILE SUCTION

Estimated average reduction: Up to 98 percent
Cost of installation: $1,200
Yearly operating cost: $15 for fan electricity, $125 for supplemental heating

Does your house have a drain-tile system surrounding it? You may not even know. The person you *bought* the house from might not even know. The person who built it will, though.

Drainage tiles, actually plastic or clay drain pipes, are sometimes installed during construction right against the foundation footing of a house. Some drain-tile systems may loop around the entire house; others surround only part of the foundation. The purpose of these pipes is to collect groundwater and route it away from the basement walls, where it could soak through and cause damage. If the ground has a natural slope, the water may drain to an above-grade "soakaway." If not, it may go to a sump pump in the basement.

Radon researchers have found another use for these drain loops. They know that a significant amount of radon gets into some houses through the joint between basement walls and the foundation footing. What would happen, they wondered, if a fan were used to pull a slight vacuum into a drainage tile system around the footing?

What happened was exactly what they hoped would: When the system was depressurized, it sucked radon out of the surrounding soil and into the pipe loop. From there, it was a simple matter to design modifications to a drain-tile system that would transform it into a neat system for collecting radon gas from the soil around a house and venting it harmlessly into the outdoor air.

So much for the good news. The bad is that if your house doesn't already have a drain-tile system that makes a complete loop around the house, it probably never will. (Check your home's blueprints, if you have them, or try to locate the contractor who built the house to find out if there is a drain-tile system around it.) The cost of installing such a system *after* a house is completed would be prohibitively expensive. And if your drainage tiles go only part of the way around the foundation, or are badly clogged or damaged, they'll be useless for radon reduction.

If, however, your drain-tile system is complete and intact, you're in business. The first thing you'll need to do is to find a radon mitigation contractor who has had some experience installing drain-tile suction systems. You can figure on estimates for installing the entire system at about $1,200. If you want to do some of the work, such as buying and installing the fan and backfilling the pipes yourself, the bill might be less.

If your system drains naturally into a soakaway, the first assignment for the workers will be to find the discharge line that leads from the drain loop to the drain exit. It's on this run of pipe (not on the loop itself) that the suction fan will be installed on the end of a 2- or 3-foot-high riser made of plastic sewer pipe. The fan and riser can be fitted anywhere on the discharge line, but the farther they are from the house, the less fan noise you'll hear.

The EPA very rarely recommends specific products for mitigation work, preferring the generic approach. They've broken their rule when it comes to fans, though. The reason, says the EPA's Gene Tucker, is that inadequate fans, and fan failure, sometimes sabotage an otherwise excellent installation.

"The fans for these systems have to pull a lot of suction," he says. "We've seen several instances where removing a cheap fan and putting in a good one dramatically changed the way a system performed. So when we ran across a fan that seemed to be able to do the job better, we got excited about it."

That fan is a centrifugal model designed in Sweden and assembled in Florida. The Kanalflakt K-series comes in 50 to 900 CFM sizes and fits handily on the end of pipes 4 to 12 inches in diameter. Suggested retail prices for all sizes range from $85 to $200. For more information and the address of your local distributor, write to the following: Kanalflakt, Inc., 1121 Lewis Avenue, Sarasota, FL 33577, or call the company at (813) 366-7505.

Whatever brand you ultimately choose, it's critical that the fan be mounted tightly on the riser, since leaks will diminish the suction on the drain loop.

Drain-Tile Ventilation Systems

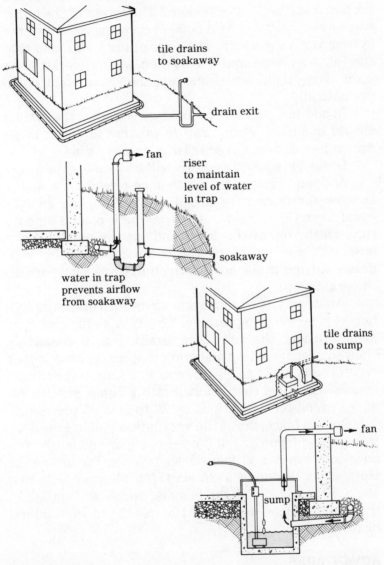

If your house has an intact drain-tile system that surrounds the entire house, you may be able to add a simple fan installation that will use the drain pipes to draw radon gas from the soil and vent it outdoors before it has a chance to get into the house.

"I remember one system we inspected that looked perfect installed," says Tucker. "The owner of the house was an engineer and he'd done everything right. But the system wasn't working. Then one of my men found that the fan was loose and the suction wasn't working. As soon as we tightened things up, the radon went down dramatically."

In addition to the riser and fan assembly, the system should include a water trap to prevent air from being drawn into it from the soakaway exhaust pipe.

Drain-tile suction systems will look quite different if the drainage system is one that empties into a sump. Designs for these setups call for building an airtight cover over the sump pit and running a plastic pipe up through the top of the cover and out through a nearby basement wall. A fan on the outdoor end of this pipe draws suction in the sump, which in turn depressurizes the drain loop.

Either way, the radon reductions with drain-tile suction can be very impressive. The EPA's study of four houses where the tile loops drained to a soakaway showed that installing a suction system lowered indoor radon levels by 74 to 98 percent. In a later test of three houses with tiles that drained into a sump, suction reduced radon concentrations by 70 to over 95 percent.

Once in place, drain-tile ventilation systems need little care and feeding. You'll need to check and oil the fan once in a while and inspect the system for broken or worn seals. Running a 25-watt fan all year will cost about $15, and the slight increase in house ventilation the system causes may add $100 to $125 to your annual heating bill.

ADVANTAGES
OF DRAIN-TILE SUCTION

- Can be installed outside, so there's no need to tear up a finished basement

- High radon reductions; very cost effective
- Homeowner may be able to do some of the work

DISADVANTAGES OF DRAIN-TILE SUCTION

- Adds to heat and electric bills
- Can be somewhat noisy
- Only works on some houses
- Must be operated continually

BLOCK WALL VENTILATION

Cost of installation: $2,500 to $5,000
Estimated average reductions: Up to 99 percent
Yearly operating costs: $15 in electricity; $125 in extra
  heating costs

If you haven't got a drain-tile system around the outside of your house to help carry away radon, you'll have to attack the problem from indoors. For houses built with basement walls of hollow concrete block and cinder block walls, one way to do that is by ventilating those walls to draw radon gas away from the house.

A couple of strategies have been developed to accomplish that. In one, a sheet metal "baseboard" duct is built around the bottom perimeter of the wall, on the inside. Holes are drilled through the row of blocks behind the duct and a fan creates suction that pulls radon-laden air out of the blocks, around the baseboard duct, and out through a pipe that exhausts outdoors.

In a simpler version, plastic pipes are inserted directly into the inside or outside of the walls and connected to a fan (or fans) that draws suction to sweep the radon out of the walls.

Either way, block wall ventilation can be a very effective cure for very serious radon problems. It's capable of cutting levels by better than 90 percent in some houses.

For any of the approaches to work its best, it's extremely important that all major cracks and openings in the walls be sealed first, and that the top row of blocks be capped, as described above in "Sealing Entry Points." If any major leaks are overlooked, the fans won't be able to pull a proper vacuum in the walls and performance of the system will suffer.

Sealing up block walls is tedious work, but you can do a fairly good job of it in most houses if you take your time. The exceptions are houses with brick veneers on the outside and some houses with fireplaces.

The problem with houses that have brick veneers is that there's usually a small space between the inside of the bricks and the outside of the block wall. Like the hollow centers of the blocks, this gap can also channel radon up from the soil level. But there's no simple way to get at it once the house is built, so even if block suction is used, the gap will continue to be a source of indoor radon.

Fireplaces cause trouble when they're set partially or entirely into the block wall. There are bound to be numerous openings between the fireplace and the blocks, all of them inaccessible. Sealing up the rest of the wall won't do much good as long as these hidden leaks sabotage the block suction.

The EPA is experimenting with ways to make wall suction applicable to all houses with basements made from hollow blocks, but in the meantime, if your house has a brick veneer or a fireplace that penetrates the wall, this method is probably going to be a disappointment. For most other houses, though, block wall suction is a straightforward, if labor intensive, job.

PIPE-IN-WALL APPROACH

A pipe-wall system is the cheapest route to take, but you'll have to live with a network of exposed plastic pipes snaking through your basement. In order for the setup to work properly, there must be at least one suction pipe in every wall that rests on footings. That would mean all perimeter walls, as well as interior basement walls, if they penetrate the slab and are supported by the footings below.

The EPA's rule of thumb is that there should be one suction pipe for every 24 feet of wall length. If, for example, you have a 24-foot wall, you'd put one pipe in the center, at the 12-foot point. For a wall longer than 24 feet, you'd want two suction points, one a quarter of the way in from each side.

The agency recommends using Schedule 40 plastic sewer pipe with a 4-inch diameter. To install each pipe in

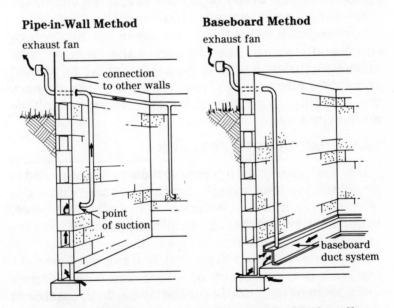

Pipe-in-Wall Method

exhaust fan

connection to other walls

point of suction

Baseboard Method

exhaust fan

baseboard duct system

Here are two methods of ventilating concrete block walls to draw out the radon that can collect in the hollow centers of the blocks. Using one technique, holes are drilled into the blocks and a baseboard duct system is installed to carry off the radon. Another approach is to imbed a pipe directly into the block cores so that air is drawn out.

the wall, you or your contractor would first chisel or drill about the same diameter as the pipe through to the center void in one block. (The closer to the floor the block is, the better the system will work.) After inserting the pipe, any gap between it and the block must be tightly sealed with an asphaltic caulk.

Once there's at least one pipe in every wall, you'll need to connect them all to the 6-inch collector pipe that will vent the collected radon gas outdoors. The best way to do that is to put an elbow on each wall pipe and run extensions up to the height of the floor joists, where they could be tapped into the larger collection pipe. This central pipe could then be run up between the floor joists, penetrate the wall, and be connected to a fan outdoors. (Another option would be simply to run the end of the collection pipe out through a basement window.)

Since you'll be venting radon gas in potentially high concentrations from the fan end of the collector pipe, make sure the exhaust is at least 6 feet from the nearest operable window so that wind won't carry the gas back indoors. Make sure, too, that the discharge won't be blocked by snow or vegetation.

BASEBOARD DUCT APPROACH

Block wall suction with baseboard ducts is neater and is less obtrusive than a pipe system. Baseboard ducts also offer more uniform ventilation than individual pipes. Unfortunately, baseboard installations cost an average of $5,000, twice as much as pipes.

But if you have a house with a finished basement you can't bear to disrupt, or a basement with a French drain around the inside perimeter of the floor, baseboard ducts make sense.

The ducts must run around the entire inside perimeter of the basement walls and along any interior block walls that rest on footings beneath the slab.

After all cracks and gaps in all the walls are sealed, ½-inch holes must be drilled into each void in every block in the row closest to the floor. (Most blocks will have two separate voids.)

The ducts can be fabricated from sheet metal, or you can use commercially made "internal channel drains." The EPA says the homemade kind offer greater flexibility for sizing and fitting the contours of basement walls.

The ducts should run around the entire perimeter, enclosing the floor/wall joint or French drain and the holes that were drilled into the blocks. Attach the sheet metal tightly to the wall and floor with masonry screws. Then seal all edges with a continuous bead of asphaltic caulk. When you are done, the whole works *must* be airtight. Be particularly careful where the ducts turn corners; these joints are often the toughest to seal properly.

To connect the duct system to a fan, tap 2-inch diameter plastic sewer pipe into the duct system (caulk the joint afterwards) at one or more places, and run the pipe through a basement window or wall to the outdoors.

A single 160-CFM or 250-CFM centrifugal fan mounted on the end of the pipe is often adequate to pull enough suction in the ducts to draw out most of the radon. If not, you'll need to run two exhaust pipes, one on each end of the house, and install a fan on each.

In houses where it's proved impossible to seal all of the cracks and leaks in the wall, more than two fans may be needed to draw enough suction. But you must be careful: Too many fans pulling at the same time could depressurize your basement enough to cause your furnace, fireplace, or wood stove to "back draft," spewing deadly carbon monoxide gas into the cellar.

One way to insure against that is to *reverse* the fans, so that they are blowing air into the block walls instead of pulling it out. The theory is that the incoming air will make the pressure inside the walls higher than the pressure of the soil gas outside. When that happens, the radon shouldn't be able to force its way in. Right now, that's all still theory. The EPA says it "currently has no data on the performance of wall ventilation systems working under pressure."

However you build it, when your system is up and running, give it a thorough going over. Walking along the walls with a smoke generator in hand will tell you a lot. If the smoke is sucked into the wall, your suction system is working. If it puffs back toward you, something is wrong. More radon tests will also verify the success of the installation.

Operating costs will average about $15 a year to run the electric fans and $140 to replace heat lost from the added ventilation in the basement.

ADVANTAGES OF BLOCK WALL SUCTION

- Very high reductions possible
- Dependable
- Most materials easily available

DISADVANTAGES OF BLOCK WALL SUCTION

- Costly
- A difficult project for homeowners to do themselves
- Fans may be noisy; system may be unsightly
- Depressurization could reduce the draft on combustion appliances
- Increases electric and heating bills

SUBSLAB VENTILATION

Estimated average reduction: 80 to 90 percent
Cost of installation: $1,000 to $2,500
Yearly operating cost: $15 for electricity, $140 in extra heating costs

One of the most thoroughly tested ways to keep radon out of houses is by subslab ventilation. This method has been used to fix hundreds of homes in the United States, Canada, and Sweden, and usually results in very large reductions in radon levels—typically 80 to 90 percent, but sometimes as much as 95 percent. Fortunately, it's also one of the cheapest of the active ventilation approaches, especially if you can do some of the work yourself.

Numerous studies have shown that in areas of high natural radon, the gas tends to accumulate underneath the basement floors or slabs of houses, sometimes reaching concentrations of 10,000 pCi/l there. As pressure builds up beneath the concrete, the gas seeps in through any available cracks, joints, or openings.

Subslab ventilation employs exhaust fans to draw radon out from beneath the floor via one or more pipes run down through the concrete to the gravel below. Installation isn't as complicated as with some other methods, and materials for the average system usually run no more than $500.

As always, there are caveats. The biggest, and hardest to overcome, is the necessity of a good layer of gravel under your basement floor or foundation slab. The spaces between the pieces of gravel make it easier for radon to move about freely under the slab, and easier for a single fan to pull it 20 feet or more underneath the concrete. If there's only hard-packed earth or, worse, clay beneath the floor, even a strong fan will only be able to create suction for a few feet around the pipe.

It's a good building practice to put down a 4- to 6-inch layer of drainage stone before pouring the foundation slab of a new house. The crushed rock insulates the floor and protects it against moisture from the ground. Too bad good practices aren't always used. Unless you were there when the house was built, finding out if you've got gravel under your slab can be a chore. The builder will know, if you can locate him. The working drawings and specifications for the house can be of help, but the fact that a drawing calls for a gravel bed doesn't guarantee one was used.

When all else fails, you can always knock an exploratory hole through the concrete and see for yourself. If there's no gravel there, there are some tricks, explained later on, that can be used to make subslab ventilation workable. But the method will never be as effective as it could be with gravel.

The second caveat: Subslab ventilation produces optimum results in houses built on a slab or in basements with poured concrete walls. When the walls are hollow block, drawing radon from beneath the slab may only take care of part of the problem, because radon could still be getting in through the voids in the walls. Sometimes it's necessary to use *both* subslab ventilation and block wall ventilation; sometimes just the subslab suction is enough to lower levels to target levels. Sealing the block walls and all openings in them will help, too.

That aside, here's how subslab systems are designed and installed: First you'll have to decide how many suction pipes you need to have in your floor. One is often enough, but in houses with large basements or big radon problems, two or more may be needed. The ball park rule is to put down one pipe for every 500 square feet of floor. For one-pipe systems, the pipe should be in the middle of the floor if at all possible. With two-pipe installations, put them equal distances from each other and the walls.

The same type of 4-inch plastic sewer pipe used for the other ventilation systems works well here. Usually all that's needed to install it is a hole through the floor that is roughly the same diameter as the pipe. If you hire a contractor, he'll undoubtedly bring along a hydraulic jackhammer for this job. If you're doing the job yourself, electric jackhammers can usually be found at rental stores.

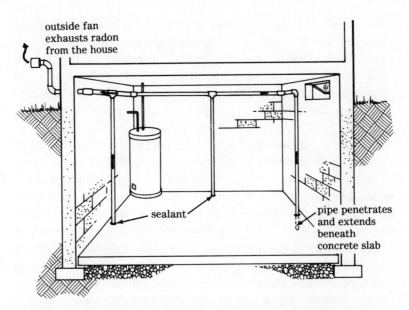

outside fan
exhausts radon
from the house

sealant

pipe penetrates
and extends
beneath
concrete slab

Subslab ventilation is one of the most common radon reme-
dies. A hole is broken through the slab and a ventilation
pipe is installed in the hole. The pipe is then run to the out-
doors, where a fan draws suction on the system and vents
radon-laden air harmlessly into the outdoor air. This system
is most effective when there is a layer of crushed stone al-
ready under the slab.

Cover the end of the pipe with a metal screen and
sink it 6 inches deep into the gravel. Then carefully seal
the edges of the broken concrete with epoxy adhesive
and fill any gaps between it and the pipe with an asphaltic sealant. As usual, it's critical to be thorough with that
part of the job.

If you're using a single pipe to ventilate your slab,
you can run it up to joist height, add an elbow, and
continue with another pipe out through the wall and up

again to a 150- to 250-CFM exhaust fan. In slab-on-grade houses, the pipe can run up through walls to the roof.

For more than one suction point, connect the pipes to a 4- or 6-inch collector pipe above with tees; then run the collector outdoors. If extra-high suction is necessary, you can also run the two pipes separately to two exhaust fans.

About those tricks to overcome the absence of a gravel bed under the floor: One method that helps is to dig a deeper and wider hole through the concrete and into the earth below and fill it with gravel. The EPA suggests excavating a hole about 1.5 feet square. Fill it with crushed rock to the level of the underside of the slab, and then install the pipe 6 inches into the gravel. A layer of building felt placed over the top of the hole at this point will prevent the mortar you'll be pouring in next from plugging up spaces between the rocks. Repair the prepared hole with concrete, and then seal and caulk the joint.

The cost of operating and maintaining a subslab ventilation system is essentially the same as for wall and drain-tile suction outfits: $15 yearly to power each fan and $140 per year to replace the heat the ventilation will rob you of.

Some experimenting is currently being done with so-called passive ventilation systems that use wind-powered turbines atop roof-level exhaust stacks, instead of an electric fan, to provide ventilation to subslab systems. The advantages of passive systems are numerous—no fan noise, no need to run wires and buy electricity—but so are the drawbacks. "There are problems with passive systems," says EPA environmental engineer Michael Osborne. "The biggest one is not being able to get large enough reductions with them. And they can be tempermental, depending on the atmospheric pressure and the weather. The system may work great one day, then a cold day comes and it doesn't work well at all. At this point, we're not recommending them."

ADVANTAGES
OF SUBSLAB VENTILATION

- Large reductions possible
- Homeowners may be able to do some work themselves
- Can sometimes be installed in finished basements

DISADVANTAGES
OF SUBSLAB VENTILATION

- Can be noisy
- Needs gravel base
- May not work well with block-wall basements
- Increases heating bills
- If the openings created are not well sealed, radon levels could increase

DEALING WITH RADIOACTIVE BUILDING MATERIALS

Houses with high radon levels caused *entirely* by building materials are rare. More often, brick, concrete containing alum shale, stone, and some gypsum wallboards give off low levels of radioactivity that add to an existing problem.

In those situations, taking care of the radon entry points into the house using one of the methods described earlier will often bring radon concentrations down to reasonable levels.

If, after testing, levels are still too high and the building materials are diagnosed as the cause, they'll have to be dealt with. And that's not always easy.

Increased ventilation always helps. As mentioned earlier, sometimes concrete or brick walls giving off radon gas can be sealed with a two-component epoxy paint or waterproof latex paint. Again, the surface will have to be carefully prepared and you'll probably have to use two or more coats to seal the masonry completely. Since even the best paints and sealers can deteriorate over time, especially when they're constantly exposed to moisture, you may have to reseal after a few years. A visual inspection may tell you when it's time to redo the job, but regular radon tests are a better indicator.

Another cure that's been suggested is building a new, tightly sealed frame wall a few inches from a wall made of radon-generating building material, and ventilating the space between the two. It's a plausible idea, but untested so far.

The last resort, which should be undertaken only after you've tried everything else and are positive about the source of the radon, is to remove the offending material. In the case of wallboard or a masonry veneer, that will be messy and rather expensive, but within the realm of the possible. Replacing walls and other structural parts of a house would probably be prohibitively expensive (in Grand Junction, Colorado, removing masonry fireplaces made with uranium mill tailings cost an average of $1,627 *per square foot*). It is fortunate that very few people will ever have to face that task to save their home.

NEW HOUSES:
RIGHT FROM THE START

"Most of the pressure on us has been to find ways to fix existing houses," says Gene Tucker. "We really don't know as much as we should about keeping radon out of new homes."

That should change. The agency has stepped up research in that direction and has been meeting with some of the nation's largest homebuilders, trying to work out voluntary guidelines for new construction in radon high-risk areas.

"Basically, builders would like to keep the cost of radon-proofing their new houses down to around $100," says Tucker. "They feel that if they have to add $1,000 or more to the cost of a house because of radon measures they're going to shut out a lot of homeowners."

The $100 goal seems unlikely, but the cost of radon protection in new construction is reasonable, adding, on average, between 0.5 and 2 percent to the total cost of building a house.

Since there has been little testing of these techniques (most of which have been developed by builders, not the government), the EPA's official policy is that they're "prudent, but unproven." Most of the approaches are based on simple common sense; knowing how radon gets into existing houses points the way toward construction techniques that minimize or eliminate those entry points. As an article in *Solar Age* magazine put it, "In all construction, the main rule is this: *Make it hard for radon to get in and easy for you to get it out if it turns up.*" Here's how:

SOLID SLABS

A solid slab, that stays solid and resists cracking, can be a fairly effective radon barrier by itself. Slabs should be laid with as few pours as possible to minimize "cold joints" that could let radon in later on. The slab should continue all the way up to the basement wall (no French drains), and all joints should be caulked with polyurethane.

Reinforcing mesh should be set in the middle of the slab thickness and nonrectangular slab shapes should be thickened to better resist cracking. Concrete poured in areas where the ground may settle and become uneven should be prestressed or poststressed.

SUBSLAB PROTECTION

Preventing radon from actually reaching the underside of a concrete slab is a worthy goal. Some builders in Pennsylvania and New Jersey are attempting to accomplish it by laying a plastic radon barrier over the gravel drainage base before pouring the floor. One product they're using is Trocal (Dynamit Nobel of America, 10 Link Drive, Rockleigh, NJ 07647), a thick PVC membrane originally developed as a moisture-proof roofing material. The contractors typically put down a level bed of crushed stone, then a layer of building felt, and then large sheets of the plastic membrane. The edges of the Trocal are overlapped by about 2 inches, then "welded"

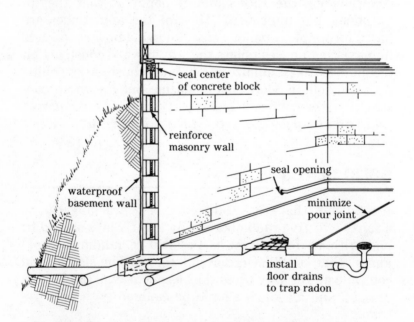

Keeping radon out of new houses is almost always easier than trying to get it out of existing buildings. Among the techniques being used to radon-proof new houses are waterproofing basement walls, avoiding too many joints when pouring concrete, sealing openings in the slab or basement walls, and using special floor drains that can trap radon.

This house is being constructed in a radon-prone area of Pennsylvania. As a precaution, thick plastic vapor barrier is being laid over a layer of gravel before pouring the foundation. The pipe stub sticking up through the vapor barrier is the exhaust port for a system of perforated pipes that will run under the foundation. If the owner ever needs a subslab ventilation system to control radon, all he'll have to do is extend the pipe and install a suction fan on its end. (Photo by Mitchell T. Mandel.)

together with a special solvent that melts a thin layer of the plastic on each sheet, creating an airtight seam after the liquid dries. After the Trocal comes a layer of sand, and then the concrete.

To play it safe, builders are also installing some extra insurance under the slab in the form of a network of perforated plastic pipes leading to a capped plastic pipe stub sticking through the basement floor. Should the house later develop a serious radon problem, everything but the fans and a few feet of exhaust pipe are in place for a subslab suction system.

RADON-PROOFING NEW HOMES

While the United States Environmental Protection Agency (EPA) is focusing a lot of attention on radon remediation methods in existing homes, the National Association of Home Builders (NAHB) National Research Center is addressing the issue of radon-proofing of new homes. The goal of the NAHB National Research Center is to keep the cost of radon-proofing a home under $100, or to attach no cost at all to the procedure. Currently, the NAHB National Research Center is field-testing 100 homes in New Jersey, in a research venture funded by the EPA, the New Jersey Builders Association, and the New Jersey Department of Community Affairs, to determine approaches that are most cost-effective in radon-proofing new homes.

This research project is field-testing the effectiveness of a two-prong approach: sealing all of the gross openings in a basement (including drains and sumps), and providing a gravel bed under the basement floor, which can be vented if radon levels turn out to be high in a new home. The venting can be done passively, with a vertical stack vent, or actively, with a small fan.

FLOOR DRAINS

A potential entry point for radon gas can be eliminated by using only floor drains with a U-shaped water trap. As long as there is water in the trap, all or most of the radon gas will be trapped behind it. If you want to be extra careful about your drains, there's a new device called The Dranjer, designed especially for use in high radon areas. Manufactured by a Canadian firm (Dranjer Corporation, 1441 Pembina Highway, Winnipeg, Manitoba R3T 2C4), the drain has a water trap *and* a ball

Researchers believe that this two-prong approach should handle the radon problems in 99 percent of all new houses. The results of this study will appear in *Interim Guidance to Radon Prevention in New Home Construction*, available from the EPA in mid-1987.

John Spears, program manager for energy and indoor air quality at the NAHB National Research Center (400 Prince Georges Center Boulevard, Upper Marlboro, MD 20772-8731; phone: 301-249-4000), answers technical questions from professional home builders regarding radon-proofing new homes. Spears reminds people that home builders are a big part of the solution to radon problems. They often attend seminars, sponsored by the NAHB, throughout the United States. In light of that, people contemplating the construction of new homes should ask their builders if they have attended such a seminar. This is an important point, since it is likely that building codes will soon contain sections on radon—and it is a good guess that the NAHB will play an important part in the formulation of such codes. And, according to Spears, it is also likely that such codes will not call for measures that add unduly high costs to the construction of a new home.

check valve behind the trap that prevents the flow of gas in the direction of the house but lets water flow out. It sells for $15 to $20.

BASEMENT WALLS

Again, poured concrete should be reinforced with wire mesh to prevent cracking. Give the exterior side of the walls a coat of cement and lime, followed by several coats of asphalt for further waterproofing.

Carefully seal the voids in the top row of masonry blocks with foam sill sealer and butyl caulk (or use solid capped blocks). The interior of the wall should get several coats. Make sure any pipe or utility penetrations are sealed with high-quality caulk.

RADON-FREE WATER

Stephen and Donna Fairbanks of Leeds, Maine, earned themselves an unwanted world's record when they tested their well water and found it contained 1.6 million pCi/l, the highest concentration ever found anywhere. They were even less thrilled with their discovery when they found that radon in their water was translated into an air reading of 2,000 pCi/l in one of their bathrooms.

It was fortunate that the Fairbankses knew whom to call: Jerry Lowry, an associate professor of civil engineering at the University of Maine. Lowry is one of a small band of radon experts in that state who are dubbed the "Radon Busters" by the *Maine Times*.

Lowry brought over one of the $900 granular activated carbon (GAC) water filters he designed and installed in the Fairbankses' basement; the airborne radon levels dropped to 3.5 pCi/l within a few days.

Radon in water is a very serious problem in some areas of the country. In houses like the Fairbankses', it's responsible for virtually 100 percent of the radon polluting the indoor air. But it takes huge amounts of radon in the water to significantly raise radon concentrations in the air, because not all of the waterborne radon is released at the same time in the house. The rule of thumb commonly used is that 10,000 pCi/l of radon in water will become about 1 pCi/l, but sometimes it can translate into as much as 10 pCi/l. In general, 20,000 to 40,000 pCi/l of radon in your well water is cause for some concern.

It is fortunate that radon isn't especially difficult to get out of water. The best, and least expensive, option for homeowners is a GAC filter like the one Lowry sells. His looks like a large oxygen tank and sells for $1,000 or less, installed along with the necessary hydroneumatic pressure tank.

The tank, available in different sizes, is filled with several cubic feet of carbon, a material that readily absorbs radon. As well water passes through the filter on the way to the household supply lines, it flows through the carbon and 90 to 100 percent of the radon is absorbed. None of the carbon actually gets used up in the process, so its life expectancy may be as high as ten years.

The radon collected in the carbon doesn't exactly disappear, either. It continues to decay inside the tank, eventually forming radioactive lead-210. The tank shouldn't become highly radioactive, although Lowry cautions his clients, "Don't go hugging your carbon unit." In fact, he recommends that it be installed in the basement, and that a fence be installed around it to keep the family at least 4 feet away.

For more information on GAC water filters, contact the following:

Professor Jerry Lowry
Department of Civil Engineering
451 Aubert Hall
University of Maine at Orono
Orono, ME 04669
(207) 581-1220

Smaller, under-the-sink carbon filters have been available for years. Designed to filter the water to one supply line—a faucet or ice maker, for example—they have been marketed as water purifiers that can remove a number of common pollutants.

No company is presently advertising one of these small filters as a radon solution, but in late 1986, *Rodale's Practical Homeowner* magazine tested ten of the top-selling models, ranging in price from $150 to $300, to see how effective they would be at removing radon from water.

As a testing "laboratory," the magazine's technical staff used a house in Easton, Pennsylvania, with well water that averaged 110,000 pCi/l. They hooked each filter up individually and measured the radon content of the water before and after filtering.

According to *Practical Homeowner* technical consultant Dhiren Mehta, all of the units were able to remove an average of more than 60 percent of the radon in the first 250 gallons run through them. After that, the carbon became saturated with radon and removal efficiency dropped 10 to 20 percent. Only after some of the radon in the filters had a chance to decay—and that can take three or four days—did efficiency improve again.

Comments Mehta, "I think if someone had one or two of these filters in a house with less radon in its water than the one we used, it could be used to treat a particular water supply, such as a shower or washing machine."

"You wouldn't ordinarily reach usage of 250 gallons taking showers or doing a load of wash," he explains. "So saturation shouldn't be a problem. But if you can reduce the radon in the water from those two sources, there may be a significant reduction in the radon in your air."

FOLLOWING UP
ON RADON REMEDIES

Whether you're treating air or water, using subslab ventilation or just sealing up some cracks, you can't just fix radon and forget it. If you do, it could come back to haunt you.

After all the time and work that goes into a radon remedy, it only makes sense to check it once in a while to see if things are still performing as expected. As long as you live in the house, radon will continue to be a *potential* problem. That means you'll want to visually inspect your system as often as possible, to make sure fans are whirling, caulks and sealants are holding up well, ducts are holding together, and every piece of the project is working like it's supposed to.

After the initial work to reduce radon levels is finished, check these reductions with a short-term tester, and install an alpha-track detector to get a year-long average of your reductions. It's probably wise to run a long-term test yearly, to help you keep track of the system's performance.

And remember, if you have *any* questions before, during, or after your radon reduction work, get on the phone and start calling state and federal experts until you get a satisfactory answer.

8

THE GOVERNMENT TAKES A STAND

To watch the federal government in action against radon today is impressive. The Environmental Protection Agency (EPA) is everywhere: experimenting with remedial techniques on houses in the Reading Prong; publishing technical manuals on mitigation techniques; offering free informational booklets to homeowners, and seminars to contractors; spending money on surveys and studies. The Department of Energy (DOE) and the General Accounting Office (GAO) are also hard at work on radon publications and tests.

There's only one thing wrong with this picture: It's the sense of immediacy about these government projects that leaves the impression that radon is something awful that came out of nowhere, which the agencies are doing everything they can to find out more about. They act as if they'd never heard of radon until 1984.

That's simply not true. The United States government and some state governments have known for more

192

than a decade that radon is a contaminant in some
homes, and there were strong indications of potential
problems with indoor radon as far back as the 1950s.
Until public outcry jolted it into action a few years ago,
the history of governmental involvement with radon was
one of missed opportunities, ignored warnings, and
aborted attempts at meaningful research.

The official story, as implausible as it sometimes
seems now, is that radon, and all other types of indoor
pollution, simply fell through the cracks of the government's public health machinery. The facts imply a more
complicated story.

WHAT THE GOVERNMENT KNEW, AND WHEN

By the time the United States began building atomic
bombs in the 1940s, it had already been fairly well established that the deadly disease that had plagued uranium
miners since the 1600s—lung cancer—was caused by
high levels of radioactive radon gas in the mines.

The push was on for more bombs and nuclear reac-
tors, however, and uranium was needed to fuel them.
The uranium industry boomed and soon hundreds of
miners in the western United States were working underground, pulling the radioactive ore out of the earth.

Despite the overwhelming evidence of the dangers of
radon, the government and the mining companies ne-
glected to mention it to the miners. In the excellent book,
*Killing Our Own: The Disaster of America's Experience
with Atomic Radiation*, authors Harvey Wasserman and
Norman Solomon report on the 1979 Senate hearings on
low-level radiation, at which some former uranium min-

ers testified. One miner told of a game he and his coworkers used to play after finishing work for the day: The men would blow into a Geiger counter and see who could register the most radioactivity. "Sometimes we could put it clear off scale," the miner said. "We were not concerned that there was anything wrong."

In 1941 the National Council on Radiation Protection recommended mine worker exposure standards. At that time, the Atomic Energy Commission (AEC) was responsible for the mine conditions, which included monitoring worker exposure to radiation. Yet, whereas the government was willing to spend billions of dollars to develop atomic weapons, the AEC claimed it lacked funding to enforce mine safety. The job was left up to the states and the mine companies.

Needless to say, the mining firms did little, and the states were reluctant to get involved during an era when uranium mining not only generated plenty of tax revenue but was touted as vital to national security as well.

In 1967, according to Wasserman and Solomon, a former AEC official named Merrill Eisenbud helped to develop a device that could identify miners who had been exposed to heavy radon doses, so that they could seek early medical treatment. But both the AEC and the Public Health Service refused to use the machine, on the grounds that there was no funding available for a test program.

At about the same time, it was becoming clear that lung cancer was beginning to victimize the new generation of uranium miners. There was concern that there might soon be a flood of compensation claims that could cost the states and the mining industry millions. Congressional hearings were called and continued throughout the 1960s and 1970s. Despite one choice bit of testimony, a Joint Committee on Atomic Energy claim that it was cigarette smoking, not radon, that was causing the epidemic of cancer, federal standards for radon gas levels in uranium mines were finally established in 1971.

RADON MOVES INDOORS

In 1966, tests began to show radon problems in some buildings in the uranium mining community of Grand Junction, Colorado. The indoor radioactivity was eventually traced to uranium mine tailings, a fine sandlike by-product of uranium mining, that was used as fill and building materials in as many as 10,000 houses, stores, and schools in Grand Junction and other mining towns.

This was a man-made problem and numerous lawsuits were initiated. Since the uranium that was being milled when the tailings were produced went almost entirely into making bombs, the federal government couldn't duck the issue. In 1978, Congress agreed to commit federal funds to remove the tailings from houses built with them in the 1950s and 1960s. Estimates at the time indicated that the cost of removal would average $15,000 per house. Under the law, the EPA was to set standards to help decide which houses would be decontaminated.

The proposed radon standard issued by the EPA during the Carter administration was based on the principle that indoor radon levels should be as close as possible to the natural "background" levels of radiation, 0.2 picocurie per liter (pCi/l), that are always present outdoors.

However, when the Reagan administration swept into the White House in 1980, things began to change. According to Robert Yuhnke, a lawyer for the Environmental Defense Fund who has been involved in the mill-tailings issue for years, Reagan's Office of Management and Budget (OMB), which controls the government's purse strings, began pressuring the EPA to adopt a cost-benefit approach to cleanup. In other words, he says, the OMB wanted a standard that balanced the health benefits with the cost of cleaning up the tailings, not necessarily a standard that would provide the most protection to everyone living in the houses.

The result of that pressure, says Yuhnke, was a final standard that reduced the number of houses that would have to be cleaned up by 25 percent. That standard was 4 pCi/l, the number that also became the EPA's recommended guideline for all homeowners in this country.

"EPA's willingness to sacrifice lives in order to reduce pressure on the budget led to a decision to accept a 1-in-65 cancer risk as an appropriate health goal," says Yuhnke. "That's a radical departure from traditional goals in the range of 1 in 1,000 to 1 in 100,000. The tragedy is that many will needlessly die of lung cancer."

WHO PROTECTS US INDOORS?

Not all radon comes from uranium mines and milling, nor, as we've found out, are people living near them the only ones who have to worry about the radioactive gas.

The warnings that "ordinary" houses could contain unsafe amounts of radon started as early as the 1950s. However, it wasn't until the 1970s, after oil prices soared and people began building tighter and tighter houses, that the extent of the problem began to emerge. Wondering what these energy-efficient houses might be bottling up indoors, researchers began sampling the air. They found some contaminants they expected—formaldehyde gas from plywood and particleboard, fumes from wood smoke and gas ranges—and one they didn't expect, radon gas. One energy-efficient house in Maryland, built to demonstrate the latest techniques in insulation and weather stripping, was found to contain radon levels far in excess of recommended guidelines.

Scientists were intrigued and more homes were tested, including some older homes that hadn't been tightened, and radon was found in some of them, as well. Studies were undertaken and researchers began looking into some mitigation techniques and ventilation strategies that might eliminate radon problems in houses.

While the evidence about the dangers of radon gas continued to mount, the EPA's stance was that its regulatory arm did not reach into people's homes; that while man-made environmental pollutants—toxic wastes, industrial emissions, water pollution—were clearly under its jurisdiction, *indoor* air quality was not.

In 1981, the National Academy of Sciences, after studying the health hazards of indoor pollution, reported to the EPA that indoor air pollution was a "serious and growing problem that can cause discomfort, illness, and even death." State officials and members of Congress began asking the agency to become more involved and finally some money was committed to looking at indoor air pollution and radon.

Also in 1981, the OMB stepped in and cut 71 employees and millions of dollars from the radon research program, effectively gutting it. "I think the OMB didn't want to open a Pandora's box," says a senior EPA official, who asked that his name not be used. "They saw that it was going to get expensive to do a lot of research and testing. Since radon wasn't a man-made problem, they decided to let it lie."

The story was repeated through much of the early 1980s: The scientific community continued to warn of the potential dangers of indoor pollution, and the EPA continued to maintain that radon was none of its business, until Congress told it differently. At one point, James Frazier, a senior staff officer at the National Research Council, publicly accused the Reagan administration of "foot dragging" on indoor air pollution.

THE SUPERFUND DEBATE

The picture didn't brighten until the fall of 1986, when renewal of Superfund legislation finally gave the EPA

the explicit authority and money to sponsor and participate in radon research and mitigation, as well as investigate other indoor pollutants. In 1980, the five-year

A VOICE IN THE WILDERNESS

Dr. Andrew Gabrysh died before seeing his warnings
about radon come true. As a young researcher working
at the Oak Ridge National Laboratory in 1953, he
found that radon could build up indoors, especially if
there was radioactivity in building materials, such as
concrete, and that levels could rise above acceptable
limits. In 1984, long retired but still concerned, he told
the Johnstown, Pennsylvania *Tribune-Democrat*,
"Over 30 years, numerous attempts were made to alert
the politicians and the medical and engineering professions of living-environment hazards due to radon.

"Seemingly, nothing has been done by either government or the engineering professions to study and
eliminate the hazard, especially in schools, where children are exposed daily."

In the 1950s, Gabrysh said, he wrote a letter to a
scholarly journal, stating, "I feel it is a disgrace to our
civilization, especially our engineering profession, that
thousands of people were unnecessarily permitted to
become, and yet remain, exposed to harmful levels of
[radon]."

Gabrysh died a few months before photographs of
the Watrases' house were splashed all over newspapers around the country. In August 1986, Joseph
Gabrysh, his brother, wrote the following to me: "I am
sure my late brother would have found much personal
satisfaction in knowing that at long last a theory he
and other scientists expounded and fought for so hard,
so long, was finally being recognized as the dangerous
cancer-causing radiation he believed it was."

Superfund bill was enacted to provide a nationwide hazardous waste cleanup program. Most of the money was used to clean up the nation's worst man-made environmental disasters—mainly illegal toxic waste dump sites. In 1985, at the end of the bill's term, a series of stop-gap measures were enacted to keep Superfund from expiring. During this time, debate raged over whether radon cleanup should be included in Superfund. Radon *did* fall under the Superfund's definition of a hazardous substance, but both Congress and the EPA interpreted the law to apply only to man-made environmental problems—not naturally occurring radon, and certainly not indoor air pollution in general.

The Reagan administration was still not interested in legislation about a naturally occurring substance and was particularly reluctant to get involved in policing the environment inside America's houses. But a group of congressmen refused to let the issue die. In the Senate, Frank Lautenberg and Bill Bradley of New Jersey, Arlen Specter of Pennsylvania, Alfonse D'Amato and Daniel Patrick Moynihan of New York, and George Mitchell and William Cohen of Maine led the way. In the House, Congressmen Bob Edgar, Don Ritter, and Gus Yatron of Pennsylvania, James Scheuer of New York, and Claudine Schneider of Rhode Island pressed for action on radon and other indoor air pollutants. It was mainly through their efforts that Congress came around on the radon issue and addressed it in the Superfund Amendments and Reauthorization Act of 1986, which was signed into law in October of the same year.

The bill gave the EPA what it says it has never had: specific authority for an indoor air research program. The legislation *requires* the agency to research the health effects and levels of hazardous contaminants and radon in indoor air, and methods to measure and reduce them. It must also conduct a demonstration program "to test methods and techniques of reducing or eliminating radon where it may threaten health."

The bill gives the EPA $5 million annually for 1987 through 1989. Two-fifths of the money in 1987 and 1988 must be reserved for the radon demonstration program.

Richard Guimond, director of criteria and standards for the EPA's Office of Radiation Programs, feels it's still up to Congress to determine how big a role the EPA will play in mitigating residential radon on a national level. In the meantime, he cautions the individual states where radon is a problem not to simply sit back and wait for federal action. "Our program is a cooperative one," he says. "We're looking for some kind of contribution from the state or local government to share in the funding and implementation of radon programs. We need that kind of commitment. We aren't going to come in and do the whole thing." He sees the EPA mainly as a facilitator. "We are going to provide all of the technical and scientific information, and we will look to the states and local health departments for the delivery of services and implementation."

It is clear from Guimond's comments and also from the relatively small sum Congress has authorized for federal radon programs that the onus is unmistakably on the states. The $5 million Superfund appropriation for *nationwide* radon action seems particularly inadequate, considering that in the state of Pennsylvania alone, $3 million has been appropriated just for low-interest loans for remedial action, while another $1.2 million has been budgeted for radon research programs. If just one state appropriates $1.2 million to clean up its own backyard, it's hard to see how $5 million will have much impact on an entire country.

While the EPA continues to work cooperatively with many states in their local radon programs, there is as yet no national radon policy. In the winter of 1986-87, EPA and ten states—Alabama, Colorado, Connecticut, Kansas, Kentucky, Michigan, Rhode Island, Tennessee, Wisconsin, and Wyoming—joined forces for statewide radon surveys. The data from these programs, as well as from similar ones under way or planned in other states, will help the EPA gather nationwide statistics even before its national radon survey—scheduled to begin in July, 1987—is completed.

That survey's primary goal is to determine what the average levels of radon are in homes across the nation. Radon measuring devices will be placed in homes chosen with the help of state agencies. The current plan calls for the detectors (probably the alpha-track type will be used) to remain in place for a year before they are collected and analyzed. The nationwide survey will also involve working with the states to help determine geological characteristics that may someday help predict areas where radon is especially high.

A third aspect of the survey involves what Guimond calls "our quality assurance program," or, officially, the EPA's "Radon Measurement Proficiency Program," designed to make sure that the companies offering radon testing services to homeowners are providing accurate results. The firms submit their measuring devices to the EPA for exposure to a known radon level, then the device is returned to the companies for calculation of the radon measurement. The results are then compared to the EPA's, and in this way, the government can confirm the accuracy of the company's tests. This program is already under way, and a list of firms that have met EPA guidelines for accuracy of measurements is available to states and individuals by request. (The current list appears in Appendix B.)

The EPA has also prepared two booklets that provide brief overviews of the radon issue—*A Citizen's Guide to Radon: What It Is and What to Do about It* and *Radon Reduction Methods: A Homeowner's Guide*. Both are available directly from the EPA or from state radiation protection offices.

Guimond says he expects that there will be plenty of data coming in as the national survey is in progress, although the final report is not due until the early 1990s.

While the survey does not include any studies of cancer rates in specific areas, Guimond says that the EPA is considering funding such a study in the future. In the meantime, research in that direction is under way in Pennsylvania and New Jersey, jointly sponsored by the states and the National Cancer Institute. Extensive studies on the health effects of radon are also being done by Bernard L. Cohen, of the University of Pittsburgh.

The EPA is also conducting training programs, called "Diagnostician and Mitigation Courses," for state and local officials and contractors interested in branching out into that area. The states act as sponsors of these seminars, which bring in instructors from federal and private agencies. According to Guimond, the EPA offered about 20 of these courses in 1985 and 1986—most of them in Pennsylvania, New York, and New Jersey.

What does the future hold for more federal involvement in the cleanup of radon? Guimond says, "We anticipate that mitigation work will be funded by homeowners, builders, and the states. At this point, there are no plans for any kind of direct funding."

In other words: Don't expect the federal government to come to your town and fix the radon problem in your house. Radon mitigation contractors around the country are complaining that they're giving lots of estimates, but many people, including some with very high concentrations in their houses, are not going through with the work. Time and again, the contractors say, the reason they're given is, "I'm going to wait and see if the government is going to step in and start fixing houses."

It's easy to see how some people might get that notion. The EPA has gotten lots of publicity for the experimental work it has done in some states. These researchers have rid hundreds of houses of radon as they tried out new and old techniques. It's important to remember, though, that these houses aren't being worked on as a favor to the owners; the buildings are field laboratories that present real-life situations for the researchers to work with.

Paying for repairs on 100, even 200, houses is one thing. Fixing 8 million is quite another. "This agency will not be providing direct cash payments for fixing radon problems," says EPA Senior Policy Advisor Gene Durman. "All you have to do is work out the number to see that it's not possible."

Indeed, if all 8 million houses estimated to be above 4 pCi/l were to be cleaned up, and if repairs cost an average of $500 per house, a conservative estimate, the final bill would total $4 *billion*.

It's unlikely the states will become involved in direct payments, either. Probably, some states with serious radon problems will follow Pennsylvania's lead in offering low-interest loans to homeowners for remedial work (see "As Pennsylvania Goes . . ." below), extensive technical help and advice to residents, and free radon testers to homeowners who live in the Reading Prong areas. "We're seeing the tip of the iceberg," says Guimond, optimistically. "We'll see more and more of these types of programs as time goes on."

Many states see it another way. Although the Superfund legislation gives the EPA more authority, and some money, to tackle the radon problem, there just aren't enough funds to provide every state with all the help it needs. The EPA says the states will have to take on a large share of the responsibility for radon studies and mitigation work. But many state officials are complaining that they simply can't afford it.

"We're operating in a vacuum—we urgently need EPA standards and guidelines," Bernard Lucey, chief of New Hampshire's Water Supply Division, told *Newsweek*. "We don't have the resources to conduct surveys or assemble massive teams of experts." His state is not alone. New Mexico had only $50,000 to fight radon in 1986; Idaho had $20,000.

HOW THE STATES ARE COPING WITH RADON

You can count yourself lucky if you happen to live in a state that has taken an aggressive stand on the radon issue and has funding to offer a range of help and services. With or without the help of Uncle Sam, some states began taking a look at indoor pollution, and at radon in particular, as early as 1975. Here's a rundown of the states that have a history of radon programs and how they've fared so far.

FLORIDA

In November 1975, the Florida Department of Health and Rehabilitative Services (now the Office of Radiation Control) and the EPA conducted radon studies in homes built on reclaimed phosphate-mining lands. Three years later, the department presented its report to the governor, while the EPA took another year to complete its analysis. Both reports concluded that the homes, most of them in Polk and Hillsborough counties, had much higher radon levels than homes in areas where phosphates weren't mined. Yet half of the homes tested with high radon levels were on land that had never been mined. It was then that researchers began to think that radon was a problem that could very well be statewide.

According to Harlan Keaton, Manager of Environmental Radiation Control, "there wasn't much that came from those reports." But they did spur the governor to create a task force, which resulted in the formation of various committees, one of which became responsible for establishing standards for residential radon levels.

In 1984, the Florida Department of Health and Rehabilitative Services developed those standards, which have since been incorporated in a rule limiting radon levels to the 0.02 working level (WL) in new homes in two counties. But the legislature has delayed implementing the rule until a statewide radon survey is completed. The survey, conducted by a private company with $1 million state funding under the auspices of the Florida Institute of Phosphate Research, is scheduled to be completed sometime in 1987.

While the rule, if ever implemented, will limit radon levels in new construction, Keaton said that there doesn't appear to be any similar legislation on the horizon for existing structures. He said that several such bills were introduced, but never made it through the legislature. He predicted that the marketplace will eventually determine how radon is dealt with in existing homes.

Florida currently has no remediation program for those homeowners who find themselves living with high radon levels and, Keaton says, no decisions will be made about future radon programs until the state survey results have been studied.

MONTANA

In 1978, homes in Butte, Montana, were tested for levels of indoor radiation. Scientists believed that the radiation might have been coming from phosphate slag, which is mined in that area. The Montana legislature appropriated $100,000 in 1979 to study the extent of radon contamination, while the EPA issued an $82,000 grant for the study, which ended in 1983. It was found that 250 homes in the Butte area had radon levels above the 0.02 WL. Since then, "basically nothing" has happened, according to Larry Lloyd, chief of the Occupational Health Bureau. Lloyd says the bureau has lost all of its funding for environmental programs, even though the state has identified many homes with high radon levels. With the economy of Montana in such a bad state, he explains, "we'll be lucky if there's a radiation program at all by July 1987."

For those homeowners in Butte, it hasn't been easy. They've had to do their own corrective action to clean up their radon-contaminated homes. Lloyd said he sees nothing on the horizon in the way of state or federal funding. In addition to Montana's financial woes, Lloyd noted that "Basically, the EPA is not too terribly interested in working with non-man-made problems. We tried for years to get them involved with radon programs out here, but they wouldn't listen."

THE PACIFIC NORTHWEST

In 1979, the Bonneville Power Administration (BPA), the largest supplier of electricity in the Pacific Northwest (and a federal agency within the DOE), started a home energy conservation program. While promoting its weatherization projects, which included adding storm windows and extra insulation to homes, the BPA discovered radon problems in the homes of its customers in Oregon, Washington, Idaho, and Montana.

Four years later, an environmental impact statement was drawn up, and in 1984 the BPA began to offer free radon monitoring in customers' homes, and set an "action level" of 5 pCi/l. Since then, the BPA has been offering installation of air-to-air heat exchangers for $1,000, a substantial discount, to reduce radon levels in homes that meet or exceed the action level.

So far, the BPA has spent more than $1 million on radon testing and research, although, according to Michael Piper, radon monitoring project manager for the BPA, the BPA's main purpose is not necessarily to *mitigate* radon in homes, but to "not make it any worse."

While the EPA is currently proposing to participate in a joint program with the BPA in Spokane, Washington, this winter, Piper said, "My personal observation is that in the northwest, the EPA has been very unhelpful." He said that whether trying to get brochures or trying to get information, "we can't get an answer from them. In fact, the EPA has come to *us* for information on radon mitigation." But, Piper says, the radon issue certainly isn't new to the federal agency. "We have people here who used to work for EPA, and they've known about radon for years," he notes.

The joint EPA/BPA Spokane project will study mitigation techniques in homes that have been identified by the BPA as having high radon levels. Piper said that the goal of the project will be to come up with a workbook of mitigation techniques using locally available materials to help homeowners keep the costs down.

WASHINGTON

The state of Washington became concerned with residential radon when the BPA first released its findings in 1979-80, according to Bob Mooney, head of the environmental radiation section of the Washington Department of Social and Health Services. The state didn't feel the need to duplicate the BPA's efforts at testing and mitigation, he says, but preferred to concentrate instead on assessing the degree of health risk that radon poses, study how best to monitor radon in homes, and determine how to evaluate the results of residential radon testing.

Mooney says that there is no state legislation or budget for the radon issue—yet. He feels that the BPA studies are providing plenty of data for the state to use, but the BPA only assesses homes with electric heat, so there's a large block of homeowners "who are on their own." For those people who choose to have their homes independently tested, there is no aid in the form of low-interest loans or other programs for mitigation, though Mooney says the state may see some help from the EPA some time in 1987.

In the meantime, the state is providing EPA booklets on radon, but with an interesting disclaimer printed on the inside front cover. It reads, in part, as follows:

> The Washington Department of Social and Health Services has agreed with the EPA to distribute this pamphlet to residents of the state. However, it is the judgment of the Department's Division of Health that the lung cancer risk estimates presented in this pamphlet are excessively high. The assessment of lung cancer risks due to radon exposure was done using past epidemiological studies that appear inaccurate and incomplete. Further studies are under way which may better define the actual risk of radon exposure. With the above caveat, this pamphlet provides information for the public to make informed decisions about their risk from radon.

In other words, says Mooney, "We don't agree with the risks the EPA is projecting—we think they're exaggerating the radon risk."

NEW YORK

Since 1980, the New York State Energy Research and Development Authority (ERDA) has been studying the indoor air quality of houses in the state. In October of that year, the *New York Times* ran a comprehensive article on the dangers to residents from radon seeping into houses from stone walls and soil. The article appeared during the final months of the Carter administration, and it spurred the Radiation Policy Council in the Executive Office of the President to ask the EPA to develop a strat-

egy for determining the extent of public exposure to radon and its health effects. The subsequent election of
Ronald Reagan ushered in an administration that was
much less environmentally concerned than the Carter
administration had been, and the council's suggestions
were never addressed. Nevertheless, the ERDA took it
upon itself to get the ball rolling on radon programs in
the state.

According to Joseph Rizzuto, program manager for
the ERDA, New York has conducted three different programs involving radon measurements since that time.
The first was a study done in cooperation with the Lawrence Berkeley Laboratory and the Rochester Gas and
Electric Company that involved homes in the city of
Rochester. The second, completed two years ago, was a
study of radon levels in homes in upstate New York. The
third is a program now taking place that involves random samplings of 2,500 homes across the state to determine radon concentrations. The EPA is contributing
$100,000 toward the study, with the state and utilities
picking up the rest of the tab.

The EPA has also been involved in two other New
York projects—a look at remedial action in existing
homes and a study of options for radon-resistant new
construction. At this time, says Rizzuto, New York offers
no financial help for residents who need to take remedial
action. However, he adds, while homeowners must finance their own radon cleanup, the ERDA is considering
providing some type of follow-up assistance.

NEW JERSEY

In 1981, aerial flyovers using radiation detectors picked
up "hot spots" over homes in three municipalities in New
Jersey's Essex County. The homes were discovered to
have elevated indoor radon levels due to radioactive

backfill materials, possibly from a company that worked with radium, that were put under the homes when they were built in the 1920s and 1930s. The findings were brought to the attention of the EPA and Centers for Disease Control, and the New Jersey Department of Environmental Protection (DEP) undertook a pilot project to remove contaminated soil from under and around 12 homes. Since this was considered a "man-induced" radon problem, the three municipalities were listed as Superfund sites, and therefore were eligible to receive funding from the EPA for cleanup.

According to George Klenk, a DEP spokesperson, New Jersey officials became concerned with the problem of *naturally* occurring radon after hearing about the Watras house in nearby Pennsylvania. In 1985, the DEP initiated contact with officials at Pennsylvania's Department of Environmental Resources to study the radon problem. In March, 1986, a man in Clinton had his home independently tested for radon. The test results showed 5 WL, 250 times higher than the EPA's guideline. The homeowner quickly contacted the DEP, and that was the start of New Jersey's current radon remediation program.

Klenk explained that the radon problem in Clinton is due to an unusual limestone formation. Homes there were built on a ridge adjacent to a former limestone quarry, and the limestone there "is extremely radioactive." The EPA has provided funding for remedial work on ten homes of various design in the Clinton area to determine the best way to clean up the radon, and the 1986 Superfund bill grants $7.5 million to New Jersey for transportation and storage of radon-contaminated soil from the houses where it was used as backfill (the barrels of radium-contaminated soil were considered toxic waste and a controversy erupted over how they should be disposed of).

In the winter of 1986–87, an independent testing firm was contracted by the state to test 6,000 homes throughout New Jersey for radon. The tests will be repeated in the summer of 1987. While New Jersey currently has a low-interest loan program for homeowners who need remedial work done on their homes, it looks as though there won't be any major funding provided to residents. "At this point (remedial action) appears to be a homeowner's responsibility," Klenk said.

PENNSYLVANIA

In 1981, the Pennsylvania Power and Light Company began an energy conservation program that included insulating homes for better heat retention. Before the program got under way, however, the company tested 36 of its employees' homes to determine what effect increased insulation would have on indoor air quality. What they discovered were unsafe radon levels in some of the homes. Further study revealed that the soil around the homes played a much larger role in radon levels than did the insulation. The Pennsylvania Power and Light Company presented a report of the findings to the Pennsylvania Department of Environmental Resources (DER) and the EPA in 1981.

Today Pennsylvania "has the largest testing program in the United States," according to Jason Gaertner, Community Relations Coordinator for the DER. But back in 1981, the state didn't take immediate action on the radon issue, primarily because it was preoccupied with the Three Mile Island nuclear accident, which occurred in March 1979, and its aftermath. It wasn't until a few years later that things calmed down and the commonwealth took another look at radon.

According to Gaertner, the commonwealth's 1985–86 budget, drawn up and submitted in 1984, included a request for a statewide investigation into radon and established the Radiation Protection Act, which provided the state's Radiation Bureau with increased funding and staff. All of which, Gaertner pointed out, happened prior to the discovery of problems at the Watras house, which brought national attention to Pennsylvania's radon troubles.

"We were the first kids on the block to come out with booklets and other information on radon because there was nothing else around," Gaertner said. After the Watras furor, the federal government became more involved and intensive radon research was begun.

The commonwealth's 1986–87 budget includes $1.2 million for radon research. In addition, there is a $3 million low-interest loan program available for homeowners whose homes need remediation work, as well as a $1 million radon gas demonstration project under way—a

AS PENNSYLVANIA GOES . . .

Pennsylvania has led the way in a number of areas concerning radon. The commonwealth has given away more than 20,000 testers to Pennsylvania's residents, produced detailed information on remedial action, and opened a hotline (800-23-RADON) to answer homeowners' questions. But perhaps its most ambitious step was a $3 million loan program to help homeowners in that state deal with radon.

Under the bill, enacted in May 1986, residents can obtain special loans of up to $7,000 to fix radon levels higher than 2 picocuries per liter (pCi/l). The amount of interest they pay on the loans depends on their incomes. Homeowners who make up to $33,375 pay 2 percent. Those with incomes higher than that pay 8.75 percent.

Local banks handle the loan applications, then forward them to the Pennsylvania Housing Finance Agency for processing and payment.

research project that is carrying out radon remediation work in 100 to 200 homes in the state.

While these are fully commonwealth-funded programs, the EPA is involved in other radon projects in Pennsylvania. There have been EPA/DER contractor training seminars, which involved instructors from the federal government as well as the private sector, and building contractors. Its primary focus, according to Gaertner, was on remedial action for eliminating radon from existing homes, though options for new construction were also discussed. There is no talk at present for instituting building codes that would require "radon-free" new construction or establish a maximum indoor radon level for new homes. And Gaertner foresees a time in the not-too-distant future when all home buyers in southeastern Pennsylvania will insist on radon tests and disclosure clauses in sales contracts before they purchase a home.

Another cooperative commonwealth/federal venture is a demonstration project called the "EPA Home Evaluation Program," which involves 80 homes in Pennsylvania and more than 100 nationwide. In Pennsylvania, the EPA is funding and contracting out to six testing firms to do diagnostic work on homes in the Reading Prong that were chosen by the DER. The program is totally funded and administered by the EPA. According to Gaertner, the EPA can provide more sophisticated diagnostic work than Pennsylvania can, and the program should result in a better understanding of the radon problem and how best to approach remediation.

OREGON

Despite the fact that the BPA was monitoring radon in homes in Oregon during the early 1980s, it wasn't until 1983 that the state itself began its own radon work. "Our program is rather superficial in comparison with Pennsylvania's and some of the other states in the Reading Prong," says George Toombs, health physicist in the Department of Radiation Control (DRC). The DRC has done some verification of BPA studies and is currently conducting statewide samplings using radon measuring devices in homes. The devices will stay in the homes for year-round sampling.

Many of the homes being tested are in the Willamette Valley, where high radon levels have been found in the past. But according to Toombs, there is no remediation program in Oregon at this time because the radon levels they've seen so far haven't justified such a program. "We've seen some high levels during the heating season," he says, "but they've dropped down in the warmer months."

The federal government hasn't gotten very involved in Oregon's radon program. According to Toombs, "The EPA has not really considered Oregon a problem, even though we've put in requests for help." Toombs said the DRC is not sure what direction its radon program is going to take in the future. Much will depend on the results of the state's sampling, as well as the results of the EPA's national survey, although, Toombs noted, Oregon is not counting on the federal government to be of much help.

When asked about what appears to be a geographical bias on the part of the EPA—with eastern states apparently receiving far more federal cooperation than the western states claim to have had—Richard Guimond responds, "We're not ignoring anybody." Citing the BPA's radon programs, he notes, "The BPA is doing extensive survey work in states such as Idaho, Washington, and Oregon," and there is no need for EPA duplication of efforts.

MAINE

Although the state of Maine has been concerned about radon in drinking water—with good reason—since the mid-1950s, "up until about 1983, we rather naively thought we understood the radon problem," says Don Hoxie, director of the division of health engineering in the Maine Department of Human Services. The state was concentrating all its radon research efforts on radon in water and not giving much thought to radon in the soil or air. But the publicity surrounding the findings in Pennsylvania's Reading Prong region spurred the state to begin looking at radon in soil and bedrock. According to Hoxie, there is currently no statewide survey planned, and no funds are available to help homeowners remediate, although there is a program in place to deal with problems of radon-contaminated water.

However, homeowners can send water samples to a public health laboratory for testing. (For more information, residents should contact: Public Health Laboratory, Department of Human Services, State House Station 12, Augusta, ME 04333.) Hoxie says the state has a "fairly substantial data base" as a result of this test program and an earlier study financed by the EPA and National Health Service.

Homeowners whose water is found to have elevated radon levels are sent a helpful booklet on the health implications of radon in water as well as in the air, and are given help in selecting the right remedial action.

To date, researchers at the Univeristy of Maine have concluded that the risk of getting cancer from ingesting radon-contaminated water is considered low. While there are currently no federal regulations limiting the amount of radon in water, the EPA plans to have rules for water suppliers by 1988. But there will be no federal guidelines regulating water that comes to residents from wells or from small water systems that serve less than 25 customers, and they are the people most likely to have problems.

While Maine's radon-in-water program has been well tended, its radon-in-soil program is still in its infancy. Hoxie said he knows of no plans for Maine to be included

in the EPA's national survey, and that no state monies
have been appropriated to allow the state to do its own
survey. He said that at one time the EPA planned to test
indoor air samples for free, with the state installing the
radon monitors in homes and then submitting them to the
EPA for results. But, he said, "[the EPA] changed its
mind" and the tests were never conducted.

DECIDING WHO'S RESPONSIBLE

As the states struggle along, looking for answers as well
as funds, the federal government stands firm in its belief
that what is needed now are tests, surveys, and more
tests, and that it is primarily the responsibility of the
states to orchestrate radon remediation efforts, even
though it is quite clear that radon is not just a problem in
the Reading Prong, but a nationwide one. And the question remains—why isn't a national remediation effort
under way? There is overwhelming evidence that indoor
radon pollution is a serious health threat, yet for citizens
of the United States, concern over their health and wel-
fare rests largely with the luck of the draw.

From the preceding look at statewide radon programs, it's abundantly clear that *where* you live deter-
mines how much help and information you get on radon.
Is it fair to the residents of Montana that because their
state is economically pressed, their health is at risk? And
what about the residents of the state of Washington?
Their public health officials are pointedly downplaying
the health risks associated with radon. The federal government's determination to let each state handle radon
as it sees fit means that many Americans are being short-
changed in terms of information and action. And it is
clear that many states plan no action on radon in the

foreseeable future, preferring instead to sit back and wait for the federal government to take charge.

The 1986 Superfund legislation is a start. At the very least, it got the EPA involved in the problem of indoor air pollution. But while tests, studies, and surveys are all helpful, what's really needed right now is *action*. The United States is not the first country to face the issues raised by the threat of a nationwide pollutant. Both Sweden and Canada have discovered and dealt with radon on a nationwide basis, and their governments' roles in remediation are held up by many researchers, scientists, citizens, and lawmakers as examples that our own government would do well to follow.

HOW SWEDEN SOLVED ITS RADON PROBLEM

When you take a look at what the Swedish government has done about radon, and how quickly it got done, it makes the United States appear to be moving in slow motion. Time and again, when independent radon experts are asked, "What more can be done?" the answer is, "Look at Sweden."

Sweden discovered its nationwide radon problem in the mid-1960s. Researchers thought the source was a lightweight concrete (made with alum shale). The concrete was banned in 1974, but by 1979 the government realized that radon in homes was also coming from the soil. A radon commission was quickly formed. Building codes were implemented in 1980 that prohibit new houses from having radon levels above 0.02 WL. Renovated homes may not have radon levels above 0.05 working levels and existing homes may not have radon levels above 0.11 working levels. By 1982 thousands of homes had been tested for radon at the government's expense, and remediation work, at an average of $1,000 per house, was well under way.

Many people in this country point to Sweden as an example of a country that not only acted quickly to solve its radon problem but also enacted nationwide regulations—something the government of the United States has been reluctant to do.

WHAT CANADA HAS DONE

Canada has also been dealing with radon in its houses far longer than the United States has. In 1975, while the government was cleaning up radioactive sites that were the results of uranium mining, it discovered large amounts of naturally occurring radon. A national survey of 10,000 homes was undertaken and found that many of the homes had high radon concentrations. Research began on ways to block the flow of radon into homes. The result was a variety of remediation alternatives, many of which have been models for cleanup work done in the United States.

In addition to remediation work, an epidemiological study is being done by the National Cancer Institute of Canada to determine if high radon levels can be positively linked to lung cancer.

MANY VOICES MAKE CHANGE

As we've seen, until recently, the United States government was doing a pretty good job of *not* publicizing information about the health implications of radon exposure. But now that the cat is out of the bag, what does the future have in store? The EPA's answer, at this point, is that it is waiting for the results of its nationwide survey—information that won't be officially tabulated until sometime in the early 1990s. In the meantime, the agency is spending money and manpower helping states conduct their own surveys and remediation efforts. But the clincher here is that the EPA only helps those who help themselves. For economically depressed states, with no hope of funding their own radon programs, the EPA has little to offer.

A lot can happen in just a few short years. The presidential elections of 1988 could have a major impact on the role the federal government chooses to play in the radon drama. Perhaps a more proregulatory, proenvironmental administration will take on the responsibility to safeguard the health of *all* of the nation's citizens, not just those fortunate enough to live in states that are already out in front in the fight against radon.

As concerned citizens, we have the right to have our opinions heard on the matter, too. And, so far, those opinions have proved to be quite effective. After all, it was the vocal and persistent pressure from a small group of senators and representatives that raised the consciousness of the Congress about radon after years of deaf ears being turned to the same concerns voiced by scientists and researchers.

If you live in an area where high levels of radon have been found in houses, but in a state that has little or no money budgeted for surveys, educational efforts, or low-interest remediation loans, let your congressional representatives know. As Richard Guimond says, it is up to Congress to determine the role the EPA will play in remediation programs. Should the agency finance programs in some states and not in others? Should the federal government be allowed to shift the burden of help to the states?

In many ways, it's up to you. If you let the people who represent your state in Washington, D.C., know what's on your mind, a lot can happen.

CHAPTER

9

WHAT IS NEXT?

The discovery of radon in the Watrases' house in 1985 was the pebble dropped into the pond. The ripples started right away, spreading outward first to the immediate neighborhood, later nicknamed "Radon Heights," then to the town and the state. More houses were tested; more problems were found. Before long, the ripples were magnified into waves that reached all the way to Washington, D.C., several hundred miles away.

That's the way it's been with radon. Pebbles keep falling into ponds all over the country. Someone reads an article about radon in a newspaper or magazine and decides to send for a test kit. The results aren't good. Neighbors tell neighbors, and they tell their friends. Someone tells the newspaper, there's another article, and the checks are in the mail for more test kits.

If you had polled residents of New Jersey about five years ago regarding the hazards of radon, their responses would probably have consisted of puzzled looks.

Who knew about radon then? But when the ripples from the newspaper articles about the Watrases' house crossed the state line, a lot of people got a quick education. In 1986, when the New Jersey Department of Environmental Protection conducted a survey, it found that 80 percent of the people interviewed knew what radon is and that it can cause cancer.

Things are beginning to move rather quickly. As late as mid-1986, the federal Environmental Protection Agency (EPA) was "officially" saying that no more than 1 million homes contained enough radon to pose a health threat. Then, Americans turned on breakfast news shows one morning and heard EPA spokesmen modify the official estimate—to 8 million houses. What can we expect in 1988? Will the new official number jump to 15 million, as some experts—even some within the EPA—predict?

Radon has been here for 5 billion years. It's going to stay as long as we do, maybe longer. As a public health issue, radon isn't going to go away. It's not like the red dye in hot dogs scandal, in and out of the newspapers within a few months. You are going to be hearing a lot more about this problem. It may change the way some of us design, build, sell, and live in our houses.

Rapid change and large areas that still fall into the realm of the unknown make it hard to predict the directions in which this issue will take us. However, based on what is known, and what's already happening in some parts of the country, it's possible to make some educated guesses. The following are some of mine.

A NATIONAL RADON STANDARD

Some states are recommending their own safety standards for radon exposure; professional and trade organizations suggest different standards, and the EPA lobbies for another. All lack the force of law. The situation is confusing to tradespeople and to homeowners. How do you decide whether to fix a radon problem, and how much you should spend doing it, when you've got three or four groups giving you different numbers to work with? How does a contractor know if the house he's selling is safe?

It makes sense to adopt one number. If I were a gambler, I'd probably put my money on the EPA's "action level" of 0.02 working level (WL), which is 4 picocuries per liter (pCi/l). The word of the federal government carries a lot of weight, and the EPA talks to a lot of people through its publications and regional offices.

Even if the 0.02 WL standard takes hold, its acceptance is never likely to be unanimous. Some critics will continue to protest that the numbers the level is based on—health statistics from uranium miners—are faulty and the public is being unduly alarmed. Others, like the Environmental Defense Fund's Robert Yuhnke, will still argue that by sponsoring an acceptable level so high "the EPA has abandoned its mission to protect public health."

With the weight of the EPA's word comes a burden. Many people believe that if the EPA says it's O.K., it must be safe. We may see further research, and debate, before a true national standard emerges.

REAL ESTATE TROUBLES

If you want to see what the threat of radon can do to home sales and property values, talk to some real estate agents who work in the Reading Prong areas of Pennsylvania, New Jersey, Connecticut, and New York. One of them sold a house in a suburb of Philadelphia twice, only

The word spreads quickly when houses with high radon levels are discovered. That usually leads to more testing and a better understanding of the extent of the local problem. (Photo by T. L. Gettings.)

to see the deals fall through because the buyers tested the building and found nearly 8 pCi/l of radon in it.

"When people hear the word 'radon,' they go running for the hills," the owner of the house told the *Philadelphia Inquirer*.

In one Carnegie-Mellon University poll of real estate professionals in the area, 72 percent said radon posed a problem in their industry. More than half said they'd lost at least one sale because of radon.

Radon tests have become standard procedure for home sales in some places—as routine as checks for termites and faulty septic systems. The 21,000-member Pennsylvania Board of Realtors has even developed a handy radon disclosure form to be attached to contracts.

Some people won't even consider a house if it has any radon at all in it. Their feelings, if they can be summed up in one sentence, are, Why should I buy this house when there are others that don't have radon problems?

At least one Pennsylvania bank is reportedly requiring a radon test before it will approve a mortgage. Richard Toohey, a radon expert at the Argonne National Laboratory, predicts that within ten years no one will be able to get a federally guaranteed mortgage until the house is certified to be below a set radon level.

You can't blame people for being afraid to buy into trouble, but you've also got to sympathize with the people trying to sell the houses. Hundreds of them only learned that they had a problem because of a test conducted as part of a sales agreement. Some were in the process of moving, or had been transferred out of state, and didn't have the time to spend designing and installing remedial systems.

Harvey Sachs, cofounder of the National Indoor Environment Institute, has proposed a potential solution: radon escrow accounts. The account would be agreed on during the negotiations and established at the closing. The seller would put in $2,000 to $4,000, depending on the type of house. During the next heating season, the new owner would run a long-term radon test, and if the levels were over an agreed-upon amount, the buyer could have them lowered using money from escrow. If the test reading is lower than the one in the contract, the seller gets back his money, plus the interest it has earned.

Complicated and time-consuming? You bet. It only seems worth the trouble when your house has become an instant white elephant because some test kits brought bad news in the mail.

Whether you are selling or buying, however, make sure you only accept a reliable test, preferably a long-term one, as evidence of a home's radon concentrations.

LEGAL PROBLEMS

Late in 1986, Dr. Joel Nobel, a Gladwyne, Pennsylvania homeowner, filed a $100,000 lawsuit against his ventilation contractor, contending that the system he installed was bringing radon into the house.

Nobel's case is one of the first of what some are predicting will be an "onslaught" of radon-related legal cases. Makers of radon detectors are being warned that a faulty tester could bring about a suit for mental stress caused by worrying about the health effects of radon.

The possibilities are endless. Someone who sells a house without testing it for radon might be sued by the buyer for negligence if the house turns out to be "hot." If a homeowner runs a test, is he legally bound to disclose the results to prospective buyers? What if he doesn't and the buyer can later prove the seller knew the house was unsafe?

Contractors and builders are losing sleep over liability questions. Some building practices that were once acceptable—unventilated crawl spaces and uncapped hollow block walls—have now been shown to be risky when there's radon about. Will the present owner of the house want to sue? And what about houses being built now? Should the contractor add features like subslab ventilation pipes or plastic radon barriers to protect *himself* as well as the occupants of the house?

As the legal counsel for a home builder's association explained it to an interviewer from the Allentown, Pennsylvania *Morning Call*: "Radon is a God-made problem. You can't sue God. Contractors have the next deepest pockets." Expect to see lots of lawyers studying up on radon in the coming years.

GRASS ROOTS ACTION

In a closet off the sunny kitchen of Kay Jones's Boyertown, Pennsylvania house are cardboard boxes stuffed full of files, memos, letters, petitions, articles, and phone logs of calls to Washington, D.C.; Harrisburg, the state capital; and to reporters and scientists. Every piece of paper in those boxes is about radon.

Jones first learned her home had potential problems with radon, which she'd never heard of, from her children. They came home one day with schoolyard stories about their neighbors, Mr. and Mrs. Stanley Watras, and a house that "glowed."

She learned about radon—quickly—and found that her own house was contaminated as well. Her fight to get action and information from the state and federal government led Jones and neighbor Kathy Varady to form an ad hoc organization called Pennsylvanians Against Radon (PAR). They set out to meet other homeowners afflicted with radon problems, lobbied state officials and United States congressmen for help, and found out everything they could about what at the time seemed like an obscure subject.

Both of their houses have since been fixed as part of an EPA test program. But the women have changed the name of the group to People Against Radon and continue to dispense advice and information on radon to homeowners around the nation.

It's no accident that Pennsylvania has attacked the issue of radon square on and come up with financial help and mitigation programs that serve as models for other states. Nor is it a surprise that the EPA has so far focused much of its attention and experimentation on Pennsylvania. It's because of individuals like Jones and Varady, who took it upon themselves to shake things up in the halls of power.

In the past year, I've traveled around the country, talking with homeowners, government officials, and people in the news media about radon. And I can tell you there are still some parts of the country where powerful health and environmental figures think the whole thing is a weird joke (in parts of the West, you can still buy a ticket to spend the day in an abandoned uranium mine, soaking up the "healthful" radon fumes).

They may be right as far as their state or county is concerned. But if they're not, it may take more groups like PAR to convince them.

OUTREACH PROGRAMS

Radon doesn't discriminate between the rich and poor. It will as happily fill up a two-room shack as a Tudor mansion. But, it seems, the owner of the mansion is far more likely to find out about it.

Several studies, among them one by Dr. Bernard Cohen, a physicist at the University of Pittsburgh, have shown that more people with high incomes and college educations test their houses than do people in lower socioeconomic classes. The implication is that poorer people may be less aware of the problem, or less able to afford a private test to find out if they have radon in their houses.

Recently, the EPA has been showing a particular interest in the data from studies that break test results

down by education and income. Although no one is saying anything just yet, it could be that the government is
planning some sort of outreach program to make sure
that everyone can get information about radon, regard-
less of how much money they have.

BUILDING CODES

It has already started. The state of Florida last year
passed a rule requiring that some new homes built near
phosphate mining areas meet certain radon requirements. In Sweden, you can't build a new house until the
lot has been tested for radon. If the test is high, during
construction the contractor must follow anti-radon
guidelines that are mandated by the government.

"I'm not sure Americans are ready for the federal
government to be traveling around the country telling
them what they should do to their houses," says Richard
Guimond, of the EPA. He's probably right, but the EPA is
at least exploring the idea. It has been working with the
major building code organizations—the Council of American Building Officials, Building Officials and Code Ad-
ministrators, the Southern Building Code Congress International, and the International Conference of Building
Officials—as well as the National Association of Homebuilders and the National Association of Realtors "to see
whether those kinds of things would be appropriate."

Some people fear that a federal building code for
radon-proofing houses would be hard to get passed and
could unfairly penalize builders in areas where radon
really isn't a problem.

But on the state, and even local, level, you'll probably see more regulations put on the books, specifying
things like ventilated crawl spaces and slabs and block

walls sealed with vapor barriers in places where radon is a potential threat.

Some zoning laws may change, too. In the rolling hills around the town where the Watrases live, new houses are still going up. There are no laws saying where, and how, the houses can be built. Some community planners and health officials question the wisdom of allowing more buildings to be erected on lands there that are known to be high in radon.

There are already laws in many areas prohibiting or limiting construction on floodplains and coastal areas subject to storm damage. Ordinances that take the same type of approach to radon hot spots are not unthinkable, although they are undoubtedly a few years away.

A NEW REAL ESTATE WRINKLE

The high levels of radon found in many homes along Pennsylvania's Reading Prong region have had an enormous impact on the real estate market there, and in other areas of the country as well. Many prospective home buyers wonder if the radon levels are too high in the homes they are considering for purchase. And there's no sure way for them to get an answer other than to run tests, or to obtain results of tests the previous owners conducted.

In an effort to put all the cards on the table in home-sale deals, the Pennsylvania Association of Realtors has devised a radon disclosure form as an addendum to the agreement of sale (see page 230).

RADON DISCLOSURE ADDENDUM TO AGREEMENT OF SALE

RE: PROPERTY _____

SELLERS: _____

BUYERS: _____

DATE OF AGREEMENT _____, 19____, SETTLEMENT DATE _____, 19____, SALE PRICE $ _____

1. BUYER acknowledges receipt of notice as set forth on reverse side hereof.

2. SELLER hereby acknowledges receipt of notice as set forth on the reverse side hereof, and certifies that:

( ) The property was tested and Radon was found to be below the EPA Remedial Action Level.

( ) The property was tested and Radon was found to be above the EPA Remedial Action Level.

   ( ) In addition, I took remedial action and represent that Radon is now below the EPA Remedial Action Level.

( ) The property was tested and no Radon was found to be present.

( ) I have no knowledge concerning the presence or absence of Radon.

3. BUYER'S OPTION (Check only one)

( ) Buyer acknowledges he has the right to have the buildings inspected to determine if Radon gas/daughters is present.   BUYER waives this right and agrees to accept the property on the basis of SELLER'S certification and agrees to the release as set forth in paragraph 4 below.

( ) BUYER, at BUYER'S expense, shall within _____ days upon approval of this agreement, arrange a Kuznet method (or equivalent) test of the residential buildings on the property.

   If the inspection reveals the presence of Radon which exceeds EPA acceptable levels, the BUYER, within five (5) days of the receipt of the report, shall notify the SELLER, in writing of the BUYER'S option to:

  a. Declare this agreement NULL and VOID, at which time all deposit monies paid on account shall be returned to the BUYER.

  b. Accept the property, which action shall constitute a release as set forth in paragraph 4 below.

NOTE: There are various laboratories in Pennsylvania through which a Radon test can be arranged, usually at a cost of under $100.00.

4. RELEASE

The BUYER hereby releases, quit claims and forever discharges SELLER, SELLER'S AGENTS, SUBAGENTS, EMPLOYEES and any OFFICER or PARTNER or any one of them and any other PERSON, FIRM or CORPORATION, who may be liable by or through them, from any and all claims, losses or demands, including personal injuries, and all of the consequences thereof, where now known or not, which may arise from the presence of Radon in any building on the property.

WITNESS _____   BUYER _____ ($ ___)

WITNESS _____   BUYER _____ ($ ___)

WITNESS _____   SELLER _____ ($ ___)

AGENT _____   SELLER _____ ($ ___)

MORE RESEARCH

We've learned a lot in a relatively short time. But there's a disturbing number of questions that remain unanswered. In writing and researching this book, I ran up against walls and stumbled into voids where information should have been, time and time again. It's O.K. to say "I don't know," but it's unsettling to hear it so often, especially from so many people who *should* know. They're frustrated, as well. They want answers to give, but the money isn't always there to go after the answers.

Slowly but steadily, funding for new radon research is growing. It's still only a paltry fraction of what's being spent on other government research, but it will help find new and better testing techniques, better ways of predicting where radon is, and surer methods for fixing difficult radon problems. Additional funding will also help to develop effective and durable tests of specific materials used in mitigation work. There will also be more money for funding small-scale cancer studies.

As long as people remain interested, and keep asking questions, the answers will come.

This sample "radon disclosure form" is widely used in real-estate transactions in high radon areas of Pennsylvania and New Jersey. It guarantees home buyers a radon test, or access to the results of a previous test, before the deal is closed. If the radon concentrations are above a predetermined amount, the buyer can get his deposit back and walk away from the sale. (Source: Pennsylvania Association of Realtors.)

APPENDIX A

WHERE TO GO FOR HELP

If you need more information on whether radon problems have been identified in your area, and if so, what is being done about them, it's best to start your information search at the local level. You might start by calling the newspaper in your town to find out if the subject has gotten some coverage that you might have missed. The librarian at a nearby public or college library may also be able to help you find information.

If your local search doesn't turn anything up, widen your circle of investigation to the state and federal levels. In many states, health agencies have prepared booklets and information sheets on radon. The agencies to contact in each state are listed below.

The federal bureaucracy can be a frustrating, impersonal entity to deal with sometimes, but if you don't get what you need from state health officials, the United States Environmental Protection Agency (EPA) may have the answers you're looking for. If nothing else, you'll find the agency to be a valuable source of free booklets and pamphlets on radon, the best of which are *A Citizen's Guide to Radon* and *Radon Reduction Methods*. Contact the EPA by calling (800) 438-2472, or write the agency's headquarters at:

United States Environmental
Protection Agency

401 M Street, SW
Suite 200, North East Mall
Washington, DC 20460
(202) 475-9605

The EPA also has ten regional offices around the country. They'll have some of the agency's literature and lists of approved testing companies, and may be able to direct you to experts in specific radon problems in your area.

EPA Region 1
John F. Kennedy Federal Building
Room 2203
Boston, MA 02203
(617) 223-7210
(for Conn., Maine, Mass., N.H., R.I., and Vt.)

EPA Region 2
26 Federal Plaza
New York, NY 10278
(212) 264-2525
(N.J., N.Y., P.R., and V.I.)

EPA Region 3
841 Chestnut Street
Philadelphia, PA 19107
(215) 597-9800
(D.C., Del., Md., Pa., Va., and W.Va.)

EPA Region 4
345 Courtland Street, NE
Atlanta, GA 30365
(404) 881-4727
(Ala., Fla., Ga., Ky., Miss., N.C., S.C., and Tenn.)

EPA Region 5
230 South Dearborn Street
14th Floor
Chicago, IL 60604
(312) 353-2000
(Ill., Ind., Mich., Minn., Ohio, and Wis.)

EPA Region 6
1201 Elm Street,
Dallas, TX 75270
(214) 767-2600
(Ark., La., N.Mex., Okla., and Tex.)

EPA Region 7
726 Minnesota Avenue
Kansas City, KS 66101
(913) 236-2800
(Iowa, Kans., Mo., and Nebr.)

EPA Region 8
999 18th Street
Suite 1300
Denver, CO 80202-2413
(303) 293-1603
(Colo., Mont., N.Dak., S.Dak., Utah, and Wyo.)

EPA Region 9
215 Fremont Street
San Francisco, CA 94105
(415) 974-8071
(Ariz., Calif., Hawaii, and Nev.)

EPA Region 10
1200 Sixth Avenue
Seattle, WA 98101
(206) 442-5810
(Alaska, Idaho, Oreg., and Wash.)

State Agencies

(Toll-free hotline numbers noted where available)

Alabama

Bureau of Radiological
Health

Alabama Department of
Public Health
Room 510
State Office Building
Montgomery, AL 36130
(205) 261-5315

Alaska

Radiological Health

Department of Health and
Social Services
P.O. Box H
Juneau, AK 99811-0613
(907) 465-3019

Arizona

Arizona Radiation
Regulatory Agency

4814 South 40th Street
Phoenix, AZ 85040
(602) 255-4845

Arkansas

Arkansas Department of
Health & Emergency
Management

Division of Radiation Control
4815 West Markham
Little Rock, AR 72205
(501) 661-2301

California

Bureau of Radiological
Control

714 P Street
Sacramento, CA 95814
(916) 322-2073

Colorado

Colorado Department of
Health

4210 East 11th Avenue
Denver, CO 80220
(303) 320-8333

Connecticut

Connecticut Department of
Health Services

Toxic Hazard Section
150 Washington Street
Hartford, CT 06106
(203) 566-8167

Delaware

Delaware Division of Public
Health

Office of Radiation Control
P.O. Box 637
Dover, DE 19903
(302) 736-4731

District of Columbia

Department of Consumer and Regulatory Affairs

Service Facility Regulation Administration
Pharmaceutical & Medical Devices Control Division
614 H Street, NW
Room 1016
Washington, DC 20001
(202) 727-7190

Florida

Department of Health & Rehabilitative Service

Office of Radiation Control
1317 Winewood Boulevard
Tallahassee, FL 32301
(904) 487-1004

Department of Health & Rehabilitative Service

Office of Radiation Control
Building 18
Sunland Center
750 Silver Star Road
Orlando, FL 32818
(904) 487-1004

Georgia

Georgia Department of Natural Resources

Environmental Protection Division
Floyd Towers East, Room 1166
205 Butler Street SE
Atlanta, GA 30334
(404) 656-6905

Hawaii

Environmental Protection and Health Services Division

State Department of Health
P.O. Box 3378
Honolulu, HI 96801
(808) 548-6455

Idaho

Division of Environment

Radiation Section
450 West State
Boise, ID 83720
(208) 334-5879

Illinois

Department of Nuclear Safety

1035 Outer Park Drive
Springfield, IL 62704
(217) 546-8100

Indiana

Indiana State Board of Health

1330 West Michigan Street
P.O. Box 1964
Indianapolis, IN 46206-1964
(317) 633-0147

Iowa

Iowa Department of Public Health

Lucas Building
Des Moines, IA 50319
(515) 281-7781

Kansas

Kansas Department of
Health & Environment

Bureau of Air Quality and
Radiation Control
Forbes Field
Topeka, KS 66620
(913) 862-9360

Kentucky

Radiation Control

Cabinet for Human
Resources
275 East Main Street
Frankfort, KY 40621
(502) 564-4537

Louisiana

Louisiana Nuclear Energy
Division

P.O. Box 14690
Baton Rouge, LA 70898
(504) 925-4518

Maine

Department of Human
Services

Division of Health
Engineering
State Health Station 10
Augusta, MA 04333
(207) 289-3201

Maryland

Department of Health &
Mental Hygiene

Division of Radiation Control
201 West Preston Street (7th
Floor Mail Room)
Baltimore, MD 21201

Massachusetts

Massachusetts Department
of Public Health

Radiation Control Program
150 Tremont Street
Boston, MA 02111
(617) 727-6214

Michigan

Michigan Department of
Public Health

Division of Radiological
Health
3500 North Logan
P.O. Box 30035
Lansing, MI 48909
(517) 335-8000

Minnesota

Minnesota Department of
Health

Radiation Control
Division of Environmental
Health
717 Southeast Delaware
Street
Minneapolis, MN 55440
(612) 623-5000

Mississippi

Mississippi State Department
of Health

Division of Radiological
Health
P.O. Box 1700
Jackson, MS 39215-1700
(601) 254-6612

Missouri

Missouri Department of
Health
Bureau of Radiological
Health
1730 East Elm Street
P.O. Box 570
Jefferson City, MO 65102
(314) 751-6400

Montana

Occupational Health Bureau

Department of Health &
Environmental Services
Cogswell Building, A113
Helena, MT 59620
(406) 444-3671

Nebraska

Nebraska Department of
Health
Division of Radiological
Health
301 Centennial Mall South
P.O. Box 95007
Lincoln, NE 68509
(402) 471-2168

Nevada

Department of Conservation
and Natural Resources
Division of Environmental
Protection
Capital Complex, Nye
Building
Carson City, NV 89710
(702) 885-4360

New Hampshire

Division of Public Health
Service
Bureau of Environmental
Health
Radiological Health Program
6 Hazen Drive
Concord, NH 03301-6527
(603) 271-4585

New Jersey

Department of
Environmental Protection
Division of Environmental
Quality
Radiation Protection
Program
CN 411
Trenton, NJ 08625
(609) 530-4000

New Mexico

Environmental Improvement
Division
Radiation Protection Bureau
1190 St. Francis Drive
Harold Runnells Building,
Room N2250
P.O. Box 968
Santa Fe, NM 87503
(505) 827-0020

New York

New York State Department
of Health
Empire State Plaza
Corning Tower Building
Albany, NY 12237
(518) 474-5422

North Carolina

North Carolina Department of Human Resources

Division of Facility Services
Radiation Protection Services
701 Barbour Drive
Raleigh, NC 27602-2008
(919) 733-4283

North Dakota

North Dakota State Department of Health

Environmental Engineering
1200 Missouri Avenue
P.O. Box 5520
Bismarck, ND 58502-5520
(701) 224-2370

Ohio

Ohio State Health Department

Radiological Health
P.O. Box 11843266-0118
246 North High Street
Columbus, OH 34266-0588
(614) 466-3543

Oklahoma

Oklahoma State Department of Health

Radiation & Special Hazards Service
1000 Northeast 10th Street
P.O. Box 53551
Oklahoma City, OK 73152
(405) 271-5600

Oregon

Oregon State Health Division
Radiation Control Section
P.O. Box 231
Portland, OR 97207
(503) 229-5797

Pennsylvania

Department of Environmental Resources

Radon Monitoring Program
1100 Grosser Road
Gilbertsville, PA 19525
(800) 23-RADON

Rhode Island

Division of Occupational Health & Radiation Control

206 Cannon Building
75 Davis Street
Providence, RI 02908
(401) 277-2438

South Carolina

South Carolina Department of Health & Environmental Control

Bureau of Radiological Health
2600 Bull Street
Columbia, SC 29201
(803) 734-5000

South Dakota

South Dakota Department of Water and Natural Resources

Office of Air Quality & Solid Waste
Joe Foss Building, Room 217
Pierre, SD 57501
(605) 773-3151

Tennessee

Division of Air Pollution Control

4th Floor, Customs House
701 Broadway
Nashville, TN 37219-5403
(615) 741-4634

Texas

Texas Department of Health

Bureau of Radiation Control
1100 West 49th Street
Austin, TX 78756-3189
(512) 835-7000

Utah

Utah State Department of Health

Bureau of Radiation Control
Division of Environmental Health
288 North 1460 West
P.O. Box 16900
Salt Lake City, UT 84116
(801) 538-6734

Vermont

Vermont State Department of Health

Division of Environmental Health
P.O. Box 70
60 Main Street
Burlington, VT 05402
(802) 863-7323

Virginia

Virginia State Department of Health

Bureau of Radiation Health
Room 914
109 Governor Street
Richmond, VA 23219
(804) 786-5932

Washington

Office of Radiation Protection

Department of Social & Health Services
Mail Stop LE-13
Olympia, WA 98504
(206) 586-3311

West Virginia

West Virginia Department of Health

Industrial Hygiene Division
151 11th Avenue
South Charleston, WV 25303
(304) 348-3217

Wisconsin

Wisconsin Division of Health

Section of Radiation Protection
One West Wilson
Madison, WI 53702
(608) 266-3681

Wyoming

Radiological Health Services

Hathaway Building
Room 478
Cheyenne, WY 82002
(307) 777-7956

APPENDIX B

RADON TESTING SERVICES

Twice a year, the Environmental Protection Agency (EPA) tests the radon testers. Any company that provides radon monitoring services is welcomed into the proficiency program, but at this point, participation is voluntary. No company has to join, but no company that doesn't pass the EPA's accuracy requirements on at least the second try gets its name in the agency's *Radon/Radon Progeny Cumulative Proficiency Report*.

To make the grade, each company must have measurement and collection methods compatible with the EPA's standards. The company is also required to submit five of its detectors, which the EPA exposes to a "secret" amount of radon and then sends back to the testing company for analysis. If the number the testing company comes up with isn't within 25 percent of the EPA's after two tries, that firm isn't put on the list.

Following is a list of companies (with the names of contact persons) that have passed the proficiency exam through the end of 1986. At any time, you can get an update by writing to: EPA Office of Radiation Programs, Washington, DC 20460.

Detector types:

AT—Alpha-track
CC—Charcoal canister
CR—Continuous radon monitor
CW—Continuous working level monitor
GR—Grab radon gas sample
GW—Grab working level sampler
RP—Radon progeny integrated sampling unit

Company (Detector Type)

AAA Radon Services (CC)

P.O. Box F
1301 Wilhelm Road
Hellertown, PA 18055
Robert Grosset
(215) 838-7164

Airchek (CC)

543 King Road
P.O. Box 100
Penrose, NC 28766
B. V. Alvarez
(800) 257-2366

Air-N-Sol Corporation (AT, CC, GR, GW)

P.O. Box 437
Frenchtown, NJ 08825
Dale Johnson
(201) 996-2028

Amersham Corporation (CC)

2636 South Clearbrook Drive
Arlington Heights, IL 60005-4692
Mark A. Doruff
(312) 593-6300

Appalachian Environmental Testing, Inc. (AT, CC, CW)

Suite 326 Masonic Building
105 South Union Street
Danville, VA 24541
Christopher R. Halladay
(804) 792-1300

Atlantic Environmental, Inc. (GW)

3108 Route 10
Suite 7
Denville, NJ 07834
Richard A. Erickson
(201) 366-4660

Biomedical Toxicology Associates (CC)

P.O. Box 3568
Frederick, MD 21701
Winifred G. Palmer, Ph.D.
(301) 662-0783

Connecticut Radon & Environmental Testing Co., Inc. (CC)

1 Pheasant Lane
Westport, CT 06880
Judith Auslander, Ph.D.
(203) 226-5173

Con-Test (GW)

P.O. Box 591
East Longmeadow, MA 01028
Gary L. Ritter
(413) 525-1198

Continental Environment Co., Inc. (CC)

34 Maple Street
Summit, NJ 07901
Alving M. Natkin
(201) 277-2255

Eberline Instrument Corporation (CW)

P.O. Box 2108
Airport Road
Santa Fe, NM 87504-2108
Eric L. Geiger
(505) 471-3232

EDA Instruments, Inc. (CC)

5151 Ward Road
Wheat Ridge, CO 80033
David Lasher
(800) 654-0506

Electro Mechanical Concepts, Inc. (CC)

130 Mountaineer Lane
West Mifflin, PA 15122
Matt Kovac
(412) 276-2272

Enrad, Inc./Rad. T. & Eng., Inc. (CC)

18705-B North Frederick Road
Gaithersburg, MD 20879
Charles L. Osterberg
(301) 948-8040

Enviradon (AT, CC)

914 Rollingwood Drive
Mt. Holly, NC 28120
Brad McRee
(704) 827-1293

Environmental Consultants Associates (CC)

14 Ramapo Lane
Upper Saddle River, NJ 07458
Richard W. Goodwin, P.E.
(201) 934-9866

Environmental Radioactivity Measure (three-day integrating radon sampler)

105 Lexow Avenue
Upper Nyack, NY 10960
Katherine Ellins
(914) 353-3513

Environmental Testing & Consulting (CC, CW)

1316 Gress Street
Manville, NJ 08835
Thomas R. Reilly
(201) 722-5293

Enviroserv (CC, CW)

15 Buckley Hill Road
Morristown, NJ 07960
Donald M. Ulbrich
(201) 285-1065

Foresight Engineering (CC)

P.O. Box 621
Madison, NJ 07940
Theresa Kaufmann
(201) 377-0602

General Health Physics (GW)

7217 Lockport Place
Lorton, VA 22079
John B. Davis
(703) 550-7525

Geomet Technologies, Inc. (CC)

20251 Century Boulevard
Germantown, MD 20874
Niren L. Nagda
(301) 428-9898

Glenwood Laboratories, Inc. (AT)

3 Science Road
Glenwood, IL 60425-1579
R. Craig Yoder
(312) 755-7911

Health Physics
Associates (AT, CC)

RD 1 Box 796
Lenhartsville, PA 19534
Anthony LaMastra
(215) 756-4153

Health Physics
Associates Ltd. (CC)

3304 Commercial
Avenue
Northbrook, IL 60062
William B. Rivkin
(312) 564-3330

HouseMaster of America
(CC)

421 West Union Avenue
Bound Brook, NJ 08805
John Hendricks
(201) 469-6050

Infiltec (AT, CC, CR)

P.O. Box 1533
Falls Church, VA 22041
Dave Saum
(703) 820-7696

Maine State Department
of Human Services (CC)

Public Health
Laboratory Station 12
221 State Street
Augusta, ME 04333
Cheryl Baker
(207) 289-2727

Mar/Gat Enterprises
(CC)

874 East Northwest
Highway
Mt. Prospect, IL 60056
Ron Garbacz
(312) 577-0979

Montana Department of
Health
& Environmental
Sciences (GR)

Occupational Health
Bureau
Cogswell Building
Helena, MT 59620
Larry L. Lloyd
(406) 444-3671

National Radon Control,
Inc. (RP)

197 Meister Avenue
Box 5342
North Branch, NJ 08876
David T. Deneufville
(201) 231-0844

New Jersey State
Department
of Environmental
Protection (CC)

Bureau of
Environmental
Laboratories
380 Scotch Road
Trenton, NJ 08628
Pat Gardner
(609) 530-4100

New Jersey State
Department
of Environmental
Protection (CC)

Bureau of
Environmental Radiation
CN 411
Trenton, NJ 08625
Duncan White
(609) 530-4050

Nucleon Lectern
Associates, Inc. (AT, CC)

2919 Olney-Sandy
Spring Road
Suite D
Olney, MD 20832
Michael S. Terpilak
(301) 774-3301

O. K. Rems Corporation
(CC, GW)

174 Flock Road
Mercerville, NJ 08619
Jeffrey C. Olcott
(609) 588-9627

Overman Associates/Air
Sciences, Inc. (CC)

P.O. Box 376
702 North Lafitte Drive
Bonne Terre, MO 63628
Ralph T. Overman
(314) 562-7020

Product Analysis &
Structural Test (CC)

6800 Wales Road
Northwood, OH 43619
Timothy Wilson
(419) 691-8484

Pyramid Environmental
Systems, Inc. (CW)

30 Oak Tree Lane
Sparta, NJ 07871
Jeffrey R. Dechacon
(201) 729-9375

R.A.D. Service and
Instruments Ltd. (RP)

50 Silver Star
Boulevard, Unit 208
Scarborough, Ontario
Canada M1V 3L3
H. L. Pai, Ph.D.
(416) 298-9200

R. K. Occupational &
Environmental Analysis
(CC)

19 Burrows Avenue
Bernardsville, NJ 07924
George D. McGuinnes
(201) 766-1737

Radiation Safety
Services, Inc. (AT, GW)

1564 Ashland Avenue
Evanston, IL 60201-4070
Eli A. Port
(312) 866-7744

Radiation Service
Organization (CC)

P.O. Box 1526
711 Gorman Avenue
Laurel, MD 20707-0953
Gregory D. Smith
(301) 953-2482

Radiation Surveys Inc.
(AT, CC, GR, GW,)

1528 Hamburg Turnpike
Wayne, NJ 07470
Carolyn Davis
(201) 628-1703

Radon Alert Detection
Center (CC)

P.O. Box 323
Flourtown, PA 19031
Ulrich W. Hiesinger,
Ph.D.
(800) 345-6348
(215) 247-1997
(215) 248-0628

Radon Analysis, Inc.
(CC)

P.O. Box 561M
Fox Run Road
Stewartsville, NJ 08886
David Chippendale
(201) 479-6086

Radon Analysts (AT)
P.O. Box 509
Livingston Manor, NY
12758
Vincent Santoro
(914) 292-2277

Radon Detection
Services, Inc. (AT, CC,
CW, GR, GW)
P.O. Box 419
Ringes, NJ 08551
James G. Davidson
(201) 788-3080

Radon Detection
Systems (AT, CC)
2300 Central Avenue
Suite B-1
Boulder, CO 80301
Tim Smith
(303) 444-5253

Radon Engineering (AT,
CC, CW)
A Unit of PSI
Engineering, Inc.
One Lethbridge Plaza
P.O. Box 549
Mahwah, NJ 07430
Harvey Greenberg
(201) 529-8300

Radon Inspection
Service (AT, CC, GR,
GW)
787 East Glen Avenue
Ridgewood, NJ 07450
Robert D. Shumeyko
(201) 670-8821

Radon Measurement and
Services (AT)
13131 West Cedar Drive
Lakewood, CO 80228
R. Stanley Thompson
(303) 988-3033

Radon Research Group
(AT)
P.O. Box 1143
6 Cross Laurel Court
Germantown, MD 20874
Michael J. Myers
(301) 972-3309

Radon Safety Services,
Inc. (CW, RP)
P.O. Box 441
107 Crane Circle
New Providence, NJ
07974
Bernard M. Odelson and
Guy Agrati
(201) 665-1188

Radon Testing
Corporation of America
(CC)
12 West Main Street
Elmsford, NY 10523
Robert M. Amram
(914) 347-5010

Recon Systems, Inc. (AT, CC)

U.S. Highway 202 North
P.O. Box 460
Three Bridges, NJ 08887
Norman J. Weinstein, Ph.D.
(201) 782-5900

Retrotec USA, Inc. (CC)

6215 Morenci Trail
Indianapolis, IN 46268
Richard A. Jordan
(317) 297-1927

Reynolds Radiological Services, Inc. (CW)

RD 1, Box 271-5
Bainbridge, PA 17502
H. W. Reynolds
(717) 653-9366

Rogers & Associates Engineering Corporation (CC)

515 East 4500 South
Suite G-250
Salt Lake City, UT 84107
Kirk K. Nielson
(801) 263-1600

Ross Systems, Inc. (AT, CC, CW)

RD 2, Box 114B
Blairstown, NJ 07825
Richard Ross
(201) 428-9088

Science Management Services, Ltd. (GW)

25 Eshelman Road
Lancaster, PA 17601
D. Dwight Browning
(717) 392-1425

Scientific Analysis, Inc. (CC)

6012 Shirley Lane
Montgomery, AL 36117
John Allen Gunn
(205) 271-0643

Sherlock Home Inspectors, Inc. (CC)

Regional Processing Center
6-01 Fair Lawn Avenue
Fair Lawn, NJ 07410
Dan Schiano
(201) 796-4942

Shotwell Associates (GW)

Box 83
40 Cedar Street
Budd Lake, NJ 07828
Henry P. Shotwell
(201) 691-9037

TCS Industries (CC)

4326 Crestview Road
Harrisburg, PA 17112
Carl H. Distenfeld
(717) 657-7032

Teledyne Isotopes, Inc. (CC)

50 Van Buren Avenue
Westwood, NJ 07675
J. David Martin
(201) 664-7070

Teledyne Isotopes Midwest Lab (CC)

1509 Frontage Road
Northbrook, IL 60062
L. G. Huebner
(312) 564-0700

Terradex Corporation
(AT)
Subsidiary of Tech/Ops
Inc.
3 Science Road
Glenwood, IL 60425
Sam Taylor
(800) 528-8327

Terradynamics
Corporation (AT)
P.O. Box 1218
Herndon, VA 22071
Ned Mamula
(703) 435-3033

Theroux Engineering
(CC)
P.O. Box 4096
380 Pine Rock
Hamden, CT 06514
Dennis R. Theroux
(203) 248-9715

TMA/Eberline (GW)
P.O. Box 3874
3635 Kircher Boulevard,
NE
Albuquerque, NM 87190
Nels Johnson
(505) 345-9921

United States
Environmental
Protection Agency (GR)
Eastern Environmental
Radiation Facility
1890 Federal Drive
Montgomery, AL 36109
Sam Windham
(205) 272-3402

United States
Environmental
Protection Agency (GR)
Office of Radiation
Programs
P.O. Box 18416
Las Vegas, NV 89114
Richard D. Hopper
(702) 798-2447

United States Toxic
Substance Testing
Bureau (CC)
Eastern Field Office
1024 Cottman Avenue
Philadelphia, PA 19111
Harold Stesis
(215) 364-3428

University of Pittsburgh
(CC)
Radon Project
Pittsburgh, PA 15260
Bernard L. Cohen
(412) 624-9246

University of Texas
School of Public Health
(CC)
P.O. Box 20186
Houston, TX 77225
Howard M. Prichard
(713) 792-4421

Virginia Department of
Health (CW)
Bureau of Radiological
Health
109 Governor Street
Room 915
Richmond, VA 23219
James A. Dekrafft
(804) 786-5932

GLOSSARY

Alpha particle—An energized particle made up of two protons and two neutrons that is ejected from a radioactive atom.

Background radiation—The average radiation from all sources that is contained in outdoor air.

Brick veneer—An ornamental layer of brick that is sometimes applied over cinder block walls.

Bronchi—The two branches leading from the trachea to the lungs.

Cinder block—Molded concrete building blocks, usually rectangular with two hollow chambers in the center. They are no longer made with cinders, but the name remains.

Depressurization—A phenomenon that occurs when the air pressure outside of a building is higher than the pressure inside. Normally, buildings have higher pressure than outdoor air, but buildings may be depressurized when wood stoves, fireplaces, or furnaces draw indoor air for combustion. The slight vacuum that is created by depressurization can draw radon-laden air into a basement through cracks and openings.

DNA—The storehouse of genetic information in cells.

Dose—The amount of radiation that a person, or specific organ, is exposed to.

Drain tile—A perforated pipe used to drain water away from the foundation of a house.

Duct—A round, oval, or rectangular pipe-like passageway for moving air. A duct can be made of plastic, metal, wood, or fiberglass.

Electrons—Negatively charged particles that orbit around the nucleus of an atom.

Electrostatic precipitator—An air cleaner that removes dust particles from the air by electrically charging them so they'll be attracted to a plate with the opposite charge within the unit.

Gamma rays—Short-wave electromagnetic radiation.

Grade—The surface of the ground surrounding the house. *Below grade* means that part of the house is lower than the surface level of the soil.

Half-life—The time it takes for a radioactive substance to decay. If the half-life of a substance is two years, half its radiation will be gone after two years.

Heat exchanger—A device that transfers heat from one medium to another, such as an air-to-air heat exchanger used for heat-recovery ventilation.

Hollow block walls—Walls made of concrete blocks that have hollow spaces in the middle. The spaces can be connected either horizontally or vertically.

Ionizing radiation—Radiation capable of exerting enough energy to disrupt atoms by knocking electrons loose. This is the most dangerous type of radiation to human health.

Micron—One-millionth of a meter.

pCi/l—The abbreviation for picocuries per liter, which is the radon gas measurement unit.

Permeability—The measure of a material's resistance to diffusion through it, in this case, diffusion of radon gas.

Plate out—The tendency of radon daughter products to attach themselves to indoor surfaces such as furniture, walls, and the insides of heating ducts. Radioactive particles that plate out are essentially rendered harmless because they won't be inhaled.

Radon—A naturally occurring radioactive gas that's colorless, odorless, and tasteless. It is formed by the decay of radium, which is itself a by-product of the radioactive disintegration of uranium. Radon gas has a half-life of 3.8 days.

Radon daughter—One of the radioactive decay products of radon gas—polonium, bismuth, and lead.

Remedial action—Work undertaken to lower radon levels in a building.

Risk—The probability that something will happen, usually used to refer to something bad.

Sill or sill plate—The horizontal wooden band that rests on top of a cinder block or poured foundation and extends around the entire perimeter of the house. The floor joists, which support the floor above, rest on the sill plate.

Slab—A level layer of concrete, usually 3 to 6 inches thick, on which some houses are built. Slab-on-grade houses, as they are called, usually don't have any basement or crawl space.

Soil gas—The gases that fill the pores between soil particles. Radon gas will be present in soil gases if the soil contains radium.

Subatomic—Smaller than an atom. In other words, the parts of an atom: electrons, protons, and neutrons.

Sump—A hole in the basement floor designed to collect water. The water may drain out by itself or be removed by a sump pump.

Voids—The air spaces created within cinder block walls by the interconnecting of the holes in the middles of the blocks.

Working level (WL)—A measurement of radon decay products originally developed for uranium miners.

Working level month (WLM)—Exposure to one working level of radon for 170 hours.

BIBLIOGRAPHY

Air Pollution Control Association. *Indoor Radon.* Pittsburgh, Pa.: Air Pollution Control Association, 1986.

Bockris, J. O'M., ed. *Environmental Chemistry.* New York: Plenum Publishing Corp., 1977.

Colle, R., ed. *Radon in Buildings.* Washington, D.C.: United States Department of Commerce, 1980.

Environmental Protection Agency. *Radon Reduction Approaches for Detached Houses: Technical Guidance.* Springfield, Va.: National Technical Information Service, 1986.

_____. *Interim Guidance to Radon Prevention in New Home Construction.* Washington, D.C.: Environmental Protection Agency, 1987.

Gesell, Thomas F., et al., eds. *Natural Radiation Environment, Volumes 1 and 2.* Washington, D.C.: United States Department of Energy, 1980.

Gofman, John W. *Radiation and Human Health.* San Francisco: Sierra Club Books, 1981.

Hendee, William R. *Health Effects of Low-Level Radiation.* Norwalk, Conn.: Appleton and Lange, 1983.

Nader, Ralph, and John Abbotts. *The Menace of Atomic Energy.* New York: W. W. Norton & Co., 1977.

National Research Council Assembly of Life Sciences. *Indoor Pollutants.* Washington, D.C.: National Academy Press, 1981.

Poch, David I. *Radiation Alert*. New York: Doubleday, 1985.

Shurcliff, William A., and David Bainbridge. *Superinsulated Houses and Air to Air Heat Exchangers for Houses*. Andover, Mass.: Brick House Publishing, 1986.

United States Department of Energy. *Handbook on Indoor Air Quality: Radon*. Washington, D.C.: United States Department of Energy, 1986.

_____. *Supplemental Combustion Air to Gas-Fired House Appliances*. Washington, D.C.: United States Department of Energy.

United States Environmental Protection Agency. *A Citizen's Guide to Radon: What It Is and What to Do about It*. Washington, D.C.: United States Environmental Protection Agency, 1986.

_____. *Radon Reduction Methods: A Homeowner's Guide*. Washington, D.C.: United States Environmental Protection Agency, 1986.

_____. *Radon Reduction Techniques for Detached Houses: Technical Guidance*. Washington, D.C.: United States Environmental Protection Agency.

United States Nuclear Regulatory Commission. *Radioactivity in Consumer Products*. Springfield, Va.: National Technican Information Service, 1978.

Wasserman, Harvey, and Norman Solomon. *Killing Our Own: The Disaster of America's Experience with Atomic Radiation*. New York: Delacorte Press, 1982.

INDEX

Page numbers in italic indicate tables. Page numbers in boldface indicate illustrations and photographs.